The New Farmer's Almanac

[AGRARIAN TECHNOLOGY]

FOR THE YEAR
2015

BY THE
Greenhorns

THE NEW FARMER'S ALMANAC 2015
[AGRARIAN TECHNOLOGY]

Published by Greenhorns
www.thegreenhorns.net

GREENHORNS

(cc) **creative commons**

NON-COMMERCIAL ATTRIBUTION
(CC BY NC)
All original artworks and articles: seek
permission from contributor to reprint.
All historical content is in the public
domain.

Printed by New York Press and Graphics
in Albany, New York

ISBN 978-0-9863205-0-7

EDITOR IN CHIEF
Severine von Tscharner Fleming

LEAD & COORDINATING EDITOR
Charlie Macquarie

LAYOUT & DESIGN DIRECTION
Norah Emily Hoover

DESIGN
Nicole Lavelle

COVER & CHAPTER ILLUSTRATIONS
Brooke Budner

CONTRIBUTOR OUTREACH
Jayme Bregman

EDITORIAL TEAM
Louella Hill
Henry Tarmy
Rose Karabush
Bruce Forrester
Audrey Berman
Inés Chapela
Patrick Kiley

SPECIAL THANKS
to the whole Greenhorns team who hustle and shuffle and make it happen in a wild
and wondrous way.

THANK YOU
The Prelingers, and the Prelinger Library for the idea, content and encouragement
to pursue this project in community history and community futurism for the new
agrarian movement. Louella Hill for sniffing out and formalizing the unique "literary
cycle" of the agrarian movement. Ruthie King and The Grange Farm School (an
amazing residential farm school in Mendocino County) for inviting us to labor and
adventure in their country in Mendocino. Tom Giessel for historical inspiration and
guidance. Inés Chapela, Ann Marie Rubin, Anna Issero, Eliza Greenman, and the rest
of the Greenhorns office staff. Joseph Wolyniak for hunting down Catholic Radical
images with such fervor. Jon Magee for his help compiling and maintaining reading
lists. Doug Mosel for the car. Mom for being an optimum feline. And especially the
many dozen passionate and flexible contributors.

We Dedicate This Book
to the Youth of
Rural America
Whose Intellectual and
Spiritual Powers and Desires
Will Determine Her Future

THIS ALMANAC BELONGS TO

....

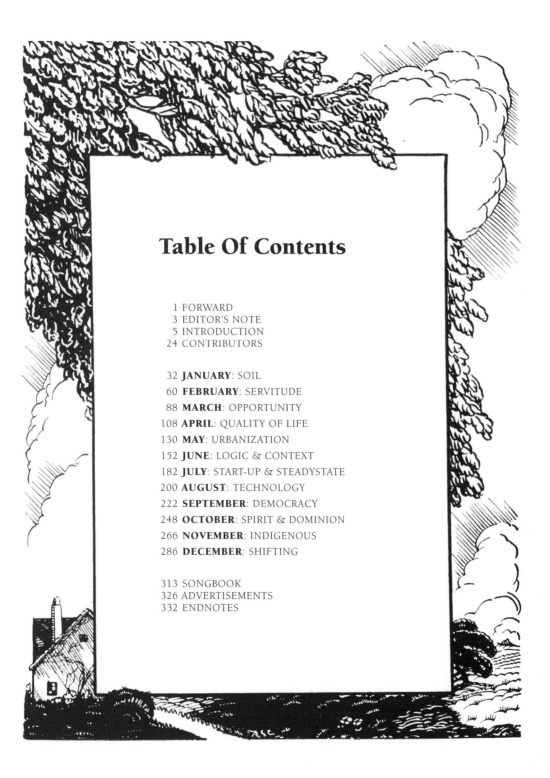

Table Of Contents

Charrue à avant-train
Polysoc
Défonceuse
Brabant double
Charrue fouilleuse
Charrue à disques
Déchausseuse
Houe à dents
Houe pleine
Charrue à disques
Extirpateur
Buttoir
Laborage à l'araire
Binette
Distributeur d'engrais
Semoir portatif
Houe mécanique
Herse triangulaire
Semoir mécanique
Semeur
Pioche à défricher
Pioche à défoncer
Herse articulée
Serfouette
Roulleau lisse
Roulleau croskill
Bêche
Rouleau ondulé
Enclumette
Faucille
Coffin
Couteau à foin
Râteau à foin
Moissonneuse-faucheuse
La fenaison
Faucheuse mécanique
Moissonneuse-lieuse
Fourche à fumier
Sape
Fourche à betteraves
Crochet de sape
Faux
Râteau mécanique
Faux armée
Fourche à faner
Arracheur de pommes de terre
Croc
Tarare
Locomobile à pétrole
Faneuse
Aéromoteur
Van
Trieur
Moissonnage mécanique
Manège à terre
Fléau
Batteuse mécanique
Batteuse à plan incliné
Motobatteuse
Locomobile à vapeur

Forward

This Almanac is our second one, but still New. The first wrinkles have arrived from squinting and grinning in the sun, but we're still young. A bit like the organization that publishes it, this Almanac is an amalgam of talents, built by a community who dip in and out of campaigns and literary works as their farm lives permit. Some members are more scarce than others—an ephemeral convening of feral cats. After the first Almanac, we decided to make it a biannual publication, which was Louella Hill's discernment at play. So let it be known that the deadline for submissions to the 2017 New Farmer's Almanac is January, 2016.

The issue currently in your hands is themed Agrarian Technology. We aim herein to present a civil, lived testimony by people whose lifeworlds and work patterns beamingly contradict the mainstream. If American capitalism is corporate canola mayonnaise, we are the wild pickle—not as a condiment, but as an inoculant.

My elder Wendy Johnson passes on bits of Buddhist wisdom whenever she's given the chance, and she recently shared some words from Thich Nhat Hanh for when things get nasty, murky, sticky and unlivable. He says "throw some straw on that mud, so we can walk over." It is a barnyard allegory, suitable for more than mud.

And the barnyard is more than an allegory—it's the poetic and powerful locus for much of the work that needs doing. It's no surprise that we're all here working in agriculture. Bent over in the sunshine, ears open to the reality of birdsong and chainsaws, bearing witness to the degraded stream-beds, and foaming aquatic dust bunnies, as drainage tiles spill fertilizer runoff into the ditches. The baseline shifts for every generation, and we'll never see the ecosystem our grandfathers knew. We struggle to measure in human terms the destruction we each cause in our modern lives. In principle a prerequisite for civilization, farming seems to have become a portal for deliberate, cultural and joyous retort against the terms of our civilizational phylogeny. In this volume, new agrarians explore alternative histories and possibilities. Tapping into a deeper, more complex past—and operating in expectation of an imaginal, but plausible, feasible, deep and tempting future.

— *Severine von Tscharner Fleming*

Editor's Note

Charlie Macquarie

Early this October I woke under a Valley Oak just before sunrise. Staring up through hanging usnea lichens as the sun hit the Redwoods across the draw, I found myself thinking of my grandmother. Before I set out for Mendocino County to finish the Almanac, she reminded me, as she often does, "there's a lot of good country up there."

If you're holding this Almanac in your hands, you've got a lot of good country to consider. Aldo Leopold's account, from his own Almanac, is apt:

> There is much confusion between land and country. Land is the place where corn, gullies, and mortgages grow. Country is the personality of land, the collective harmony of its soil, life, and weather. Country knows no mortgages, no alphabetical agencies, no tobacco road; it is calmly aloof to these petty exigencies of its alleged owners.

Leopold's distinction, and the words of my grandparents, often guide me in my own investigations of landscape and engagement with place. The complexities Leopold highlights are not news to any of this volume's gracious contributors. They live and make their livings in the nuances between land and country.

As editor I've been incredibly lucky to receive dispatches from all over this country, and other countries too. I can confirm that there is indeed some good country out there. And there is definite momentum, a welcome groundswell underneath the new (and the seasoned) agrarians who are at work throughout this Almanac.

During the editing, the fact that I'm not a farmer has occasionally bothered me. I've encountered biodynamics and orchard-planting and alternative forages, but my only frame of reference is my own small garden plot—beds of Nevada dirt piled straight onto the concrete of my East Oakland backyard.

I think of Donald Culross Peattie, who gives me a sense of purpose with his own personal charge as a naturalist-writer: "I wish that I might be, for some at least, a water carrier. This is a light task, but the human thirst is very great. I understand this from the many who call, out in the fighting." From the process of compiling this Almanac, I have seen that there is thirst. There is water to quench it as well.

With the harvest drawing to a close for much of the country, that water will be freezing soon. But of course, this will not slow the stresses on our land, the increasing structural inequalities in our food systems, or the warming of our climate. The reports in this Almanac relay struggle, but they also recount building, both of community and its infrastructure. There is unmatched vigor in those doing the fighting.

Maybe most importantly, this book's contributors remind us that we can rely on each other to slake our thirst. We've got a lot of good country out there. We're only beginning to discover it.

Introduction

Severine von Tscharner Fleming

Modernity tended to look down and backwards at Agriculture—in alternating moments romanticizing and de-politicizing or romanticizing and politicizing. The former usually from the top down and the latter from the bottom up. It is fertile ground for metaphor, for direct action, for true, naked experience, and for conjecture as well. Exploring what is possible within the boundaries of one farm and one lifetime encourages such sweeping notions of potential. As more and more of us engage in agriculture, and throw down straw, seek local ownership, experience stubborn local holdouts, watch bright stars overhead through twinkling under-stories of nurture— we become radicalized by what seems possible. An elongated sense of history provides a plentiful repertoire of resistance strategies. We hope here to hack open a few avenues, and leave a few scent-traces into the less-known micro histories of our agricultural tenure on this continent, as well as those we've replaced.

And it's important because as we build upon history there is a question of attitude. Is this scar tissue on an old stump, or is it the site of potential, for regrowth, adaption. Are we doomed?

The rules that matter are the rules of life.

Life: plant, animal, insect, and estuary-complex. Life! is the meta-economy. An alliance with ecology is the strongest, because, as Wes Jackson would tell us, "nature as measure." Because agriculture happens in the context and canvas of living systems, agriculture is exactly the right place to discover space for an economy based on sounder principles. Thank goodness for organic growth, the resilience, bounce-back, and massive productive power of photosynthesis and her sister systems. Ecological agriculture gives us the benefit of autonomy, and production, to afford rebuilding with a different priority. It's not just apps and startup companies designed, like crystal patterns, as a quick fix or convenient candy rack to tempt and sideswipe a bit of revenue from the appetites of powerful ones at the top. That logic pattern is loud and animated and it seeps through mass media and hype marketing into every corner of our brain. What a particular place wants, what it can bear, that is a quieter pattern and a softer song of possibility. But this is the song of possibility that sings in the ears of our Almanac authors. We are squinting, hands on our hips, at the sagging barn to

"The survival of humans depends on the capacity of our species to maintain and preserve the plasticity of the Biosphere with all its interacting components, the human species included. Since agriculture is a production system based directly on the resources of the biosphere—soil, water and biodiversity—it provides a good example of non-sustainability brought about by the transition from traditional knowledge to fragmented traditional science. The reductionist method, born with modern science and with the aim of simplifying the study of natural systems, led to impressive progress in technology, but also to deep fragmentation in knowledge and a lack of capacity for synthesis.

"The construction of a simplified world based on single version of a few, optimal products, both living and non living, leads to the creation of a single, homogeneous society with only one culture, one ideology, one science, one technology, one model of economy and production. In other words, it means destroying the tools and the processes that have allowed the adaptation and the proliferation of humans in all areas of the planet. It also implies the destruction of cultural and biological diversity.

"Farmers across the world are re-evaluating traditional knowledge as a source of innovation, and are following their own independent paths of development as opposed to those suggested by official systems of knowledge, and are building parallel systems of knowledge, aligning themselves with non-reductionist segments of scientific research."

—from the *Manifesto on the Future of Knowledge Systems*

"The function of planning is to render actual and evident that which is potential and inevident… This action of the imagination has already been described under the term of 'psychologic conversion.' This is closely related to biologic conversion: the action of the human brain cell is akin to that of the cambium layer. The cambium is that layer of perpetual fluidity which in the tree converts the ethereal substance of carbon dioxide into the solid substance of wood and timber."

—from *The New Exploration: A philosophy of Regional Planning*, Benton MacKaye, 1928.

"But the justifications of the family farm are not merely agricultural; they are political and cultural as well. The question of the survival of the family farm and the farm family is one version of the question of who will own the country, which is, ultimately, the question of who will own the people. Shall the useable property of our country be democratically divided, or not? Shall the power of property be a democratic power, or not? If many people do not own the usable property, then they must submit to the few who do own it. They cannot eat or be sheltered of clothed except in submission. They find themselves entirely dependent on money; they will find costs always higher, and money always harder to get. To renounce the principles of democratic property, which is the basis of democratic liberty, in exchange for specious notions of efficiency or the economics of the so-called free market is a tragic folly"

—from *Home Economics*, Wendell Berry, North Point, 1987.

see which beams can bear a new wing. Where can we prop it up? Where should we rip it down for salvage?

Restoration

"Restoration agriculture," the term, came to prominence as the title of a book by a radical midwestern edible forester named Mark Shepard. I see this functioning also as "Reconfiguration agriculture"—a working system that actively reshapes the economy, as well as the ecology, in which it operates. Attentive stewards with reform on the mind, and enough wiggle room, can sometimes set up systems that buck the mainstream. And like a little check-dam, they slow the erosive force of a river swollen in the storm.

Our agricultural system seems like just such a surging river—flowing through degraded wetlands, swollen and muddy with petrochemicals. When you're scavenging waste oil from the back of a strip-mall Chinese food joint, with infected wrists from the rat-festering tanks, that's the buttonhole of industrial agriculture. It's worth protesting, worth decrying, and hard to change. It's time to go upriver.

What do the narrow, animated upriver streams look like—those streams which constitute the headwaters of a new system, a "new economy" thats so profoundly needed? An economy that acknowledges the future, the externalities, the consequences, the integrity of the system? Unfortunately, it is not being taught as a discrete set of practices aimed at steady-state. There aren't any plans written by genius committees that chart out the way. This isn't a top-down, command and control situation. There are no orders to follow, no blue print, and if there were it would be suspect anyway. John Jeavons has his system, Alan Chadwick has his system, Fukuoka has his system, but these aren't prescriptive of how land managers, operating according to ecological principles, can conjoin their micro-economies into meso-economies.

On the heroic spectrum there are examples of dynamic, proactive formats for land health which hold inside them the rich kernel of a peaceful political economy. Both of my examples started out on rented land, and through power of virtue, charisma, luck and stamina managed to translate their commitment into ownership. A 500 acre, full diet, horse-powered C.S.A in the coldest, most sparsely populated and nearly poorest corner of New York state? Only mad people could do this. There, they produce a full diet—beef, lamb, pork, poultry, dairy, vegetables, grains—all on a weekly "free choice buffet" system with on-farm distribution. Essex Farm created its own format, overcoming structural obstacles with a pioneer ferocity, and it spawned 10 new farms in the surrounding area. Another example is Brookford Farm, with 25 milk cows, silos, wheat, pigs, cheese, milling, markets—all built knee deep, like Russian peasants in the mud of a barn-yard they did not own. They stored their grain in an old paper mill, convinced a town and millionaire to back them on a farm of their own, and built a small army of utterly devoted customers throughout the state. They even started their own bottling plant, all with four tiny children hanging off their fronts and backs.[1]

Through sheer force of will it seems, such pioneers make a new format possible in an inhospitable world. And what follows their success is a set of dynamic successional ripples—the talent that they attract, inspire, and spit out into the world. There are now 13 new farms in the towns around Essex—mostly connected in some way to the energy of initiation at Essex. The model that they invent is then observed, adapted, modified, re-enacted in other places and other towns. There is a heterogeneous pace and a very specific place-based quality to these remarkable farms, their ability to capitalize on the specific local opportunities and make a go of it.

And yes, the macro-economic conditions are inhospitable. And yes, labor is expensive and yet too cheap, food prices are higher than people can afford and yet lower than a reasonable cost of production. The technology platforms we use can connect us with markets, but also siphon away mental energy, time and resources. It is a tight turning radius to work within—to have enough yeoman, to be punk. To have enough capital to take risks. To be wily enough to adapt to change. To be sound-footed enough to dance around broken barbed wire, backing out broken equipment in a soggy autumn hedgerow.

There's a kind of natural law, obvious to people who take care of animals. It goes like this: If you are the only one who knows that the horse has no water, it's your responsibility to clean, plug, and fill the leaking trough. I'm sure we are not the only ones who have identified the massive shift, the next time coming, in agriculture, but that doesn't remove our responsibility to begin filling the trough.

It's up to us to navigate a truce making with the outgoing generation. We cannot afford to lose the land, we cannot afford to let the barn slip down, let the greenhouses lurking by the side of the road be bulldozed for a strip mall. Bankruptcy is a financial construct—ecological insolvency leaves us nowhere. We need to negotiate. We need to figure out real needs, and how reasonably they can be met. We need to assert. We need to bravely and with open-hearted, hard working honesty engage in a discourse of possibility.

Land Servitude, Farm Servitude
Hardly any agricultural societies managed without servitude. We're taught in school that hierarchy is necessary in order to maintain the irrigation, water works and infrastructures. After going to college we're reminded frequently that we should do "policy work" and other meta-labor, instead of participating directly in the stoop labor we extract from the 70% of farm workers who lack citizenship in the country they live.

Vinoba Bhave, known as the walking saint of India, marched more than 15,000 miles, mostly barefoot. Through a spiritual discourse and by applying the principles of Gandhian economics, he managed to convince many thousand landlords to gift their land into village ownership, what is known as the Bhoodan movement. By 1951, more than 4 million acres of land had been re-allocated, usually to landless farmworkers.

"We do not aim at doing mere works of kindness, but at building a kingdom of kindness.

"Money should be but an appendix on the book of life. But today, it is the sole theme of every chapter.

"The earth is the lord's and the fullness thereof' and though the land may be held in trust by individual owners, the village itself is the real owner of it, deciding how the acreage can best be divided. Vinoba wants each village to become as self-sufficient as possible in food and clothing in itself, not exporting either of these primary necessities until local needs have been met."

—from *India's Walking Saint*, Hallam Tennyson, Doubleday, 1955.

---

1   To learn more about these farms read *The Dirty Life,* by Kristen Kimball and watch *Brookford Almanac,* a film by Cozette Russell. While you're at it, watch all the films in the UP UP FARM film festival: http://www.thegreenhorns.net/up-up-farm-film-festival/

Joe Studwell produces compelling answers to two of the greatest questions in development economics: How did countries like Japan, Taiwan, South Korea, and China achieve sustained, high growth? And why have so few other countries managed to do so? His answers come in the form of a simple—and yet hard to execute—formula: (1) create conditions for small farmers to thrive, (2) use the proceeds from agricultural surpluses to build a manufacturing base focused on exports, and (3) nurture both these sectors with financial institutions closely controlled by the government.

According to Bill Gates, "The agriculture section of the book was particularly insightful. It provided ample food for thought for me as well as the whole Agriculture team at our foundation. And it left us thinking about whether parts of the Asian model can apply in Africa."

How Asia Works (www.gatesnotes.com/Books/How-Asia-Works), by Joe Studwell

Is agrarianism the passageway of empire? Is it the invisible petticoat holding up the damask bodice of unreasonable power?

Recently I organized a speaking engagement with Eric Freyfogle. An author of multiple books on agrarianism, law, nature and power, he insightfully breaks down our ideas about private property, the configuration of systems, and the whole notion of how we use nature. We have to ask fundamental questions. What does it mean to own land and other parts of nature?

Private property ownership is an arrangement of power—not just power over nature, but also power over other people. All our power is derived from nature. Behind every corporation, behind every army is a mountain, a mine, a living or geological system which we dismantle and reassemble to suit our purposes. In the USA, we have removed more than 200 mountains in my lifetime, all to make the power grid which lights up my computer screen and yours.

But even tech billionaire and biotech promoter Bill Gates, who has funded, to the tune of hundreds of millions of dollars, the "next green revolution" in Africa—has discovered that farmers are the base layer.

And our clever systems have wrought some marvels indeed. But ultimately we have to face the music: a system that reduces complexity, diversity, resilience, and spongy-bounce from its underlying ecology is a parasite. It bears the moral burden of its own embedded destructiveness.

Much of the technology we enjoy—whose "techno-logic" seems to propel itself forward according to an almost evolutionary compulsion—is all predicated on parasitizing a pre-existing ecosystem, namely our planet. Those neato silicon chips, tennis shoes, and breathable fabrics, are made by young Chinese children living in cold warehouse dormitories, with inadequate plumbing. Sent away from small farms, made marginal by macro-economics, they struggle to earn a wage.

It feels like a perpetual ponzi-scheme played out against the future. The silicon boom town covers the valley floor, replacing the cherry plantation. Those dust-blown Oakies, if they survived the Hoover towns, became soil miners themselves. Sons of gold miners, fathers of golf course sod growers, topsoil merchants, concrete quarriers, suburb curb pourers. It's one generation of extraction extracting from the next.

And this is our empire. But if agriculture is the portal into empire, can it also be a portal out of empire? Whats the relationship between the peasant and the king? The hedge funder and the stone mason? The techno startup prince, and his land manager? Can the emergent and the residual coexist?

We must remember that in this long tumble towards our current state of stratification, there have been exceptions to the rule of dog-eat-dog. There's a nice metaphor for this, given to us once again by observation of long-form nature. During the ice age the unstoppable glacial sheet would occasionally skip over a particularly sheltered bit of topography. A sideways valley would be protected by unconscious geometry and the surrounding mountains. This is called a "refugia", and these pleistocene remnant ecosystems became the epicenters of biodiversity, reseeding their barren surroundings as the glaciers receded.

On the Commons

Commons-based land governance abound in an oft-neglected literature of durable land-based societies. Mostly, these societies aren't visible from main roads, or flatlands, or the prime ecosystems which empires tend to snatch up one from another. But the more delicate, more marginal human habitats—steppes, plains, montane areas—require more careful, more detailed local knowledge. These places are less snatchable, less habitable by highly hierarchical societies—and therefore provide a sanctuary for indigenous and agro-ecological cultures, which as it happens are usually far more egalitarian. These cultures, with their land-race varieties, spiritual practices and strong awareness of boundaries and constraints have the skills and sensitivities to survive inside the boundaries of the habitat. These are lessons of relevance to a larger macro-human society that needs to reestablish its boundaries.

These systems persist in the highlands, marginal-lands, the back corners of our over-developed planet. We should study them. We, as agrarians and idealists, are inadequately fluent in the lessons that these places and their stewardship arrangements teach. I propose that we regrow a fluency with the many case-studies of durable, surviving, land-based cultures.

There are good, compelling examples all over the world—the land-sharing of the Hawaiian Aina, the 10,000 year corn culture of the Abenaki, the perennial water gardens and spiritual algorithms of paisley-shaped rice paddies, whose sacred economies out-perform the fanciest computer math.

Elinor Ostrom studied such "Common-Pool Resources," and identified 8 common characteristics, or "design principles," for steady-state societies which managed to persist atop their nature without expansion (conquest) or degradation (collapse):

The 8 "design principles" of stable local common pool resource management:

1. Clearly defined boundaries (effective exclusion of external un-entitled parties);
2. Rules regarding the appropriation and provision of common resources that are adapted to local conditions;
3. Collective-choice arrangements that allow most resource appropriators to participate in the decision-making process;
4. Effective monitoring by monitors who are part of or accountable to the appropriators;

"Wild rice, which had led to their advance thus far, held them back from further progress, unless, indeed, they left it behind them, for with them it was incapable of extensive cultivation... In civilization one class of people at least must have comparative leisure in which to develop short-cut methods of doing old things, of acquiring the traditions of the race, and of mastering new thoughts and methods. Such leisure is impossible with a precarious food supply. But, in spite of these facts, for barbaric people during the period of barbarism, the most princely vegetal gift which North America gave her people without toil was wild rice. They could almost defy nature's law that he who will not work shall not eat."

—from *Recovering the Sacred*, Winona LaDuke, South End Press, 2005.

Elinor Ostrom

5. A scale of graduated sanctions for resource appropriators who violate community rules;

6. Mechanisms of conflict resolution that are cheap and of easy access;

7. Self-determination of the community recognized by higher-level authorities; and

8. In the case of larger common-pool resources, organization in the form of multiple layers of nested enterprises, with small local CPRs at the base level.

"'Every man has a right to an equal share of the soil, in its original state,' and that 'everyone, by whose labor any portion of the soil has been rendered more fertile, has a right to the additional produce of that fertility, or to the value of it, and may transmit this right to other men.' In commenting on these maxims Ogilvie says 'on the first of these maxims depend freedom and prosperity of the lower ranks. On the second, the perfection of the art of agriculture and the improvement of the common stock and wealth of the community.'"

—from *The Green Rising*, W.B. Bizzell, Macmillan, 1926.

"There is a tendency in many histories to confuse together what we have here called the mechanical revolution, which was an entirely new thing in human experience arising out of the development of organized science, a new step like the invention of agriculture or the discovery of metals, with something else, quite different in its origins, something for which there was already an historical precedent, the social and financial development which is called the industrial revolution."

—H.G. Wells

I'm particularly passionate about the Seaweed Commons. Seaweed is the understory of the fishery, the substrate on which baby crabs, shrimps and the little creeping "sea insects" feed. Edible seaweed comes in three colors: brown, red and green. Humans eat dulse, wakame, kombu, alaria, kelp, nori, and others. Many countries practice oceanic aquaculture, growing shellfish, fish, shrimp and seaweed—sometimes in a stacked arrangement. These pieces of the aquacultural system rely on each other to function—the spark for establishment of the Commons and systems of sharing. But coming to a peaceful, gentle harvest of the wild commons is a more difficult but more valuable lesson for us to learn together.

Which brings us to sharing, and the character of sharing, the skillset of sharing, the consensus around what sharing means. It means different things to different people. The sharing economy—where nice strong bikes are available with the swipe of a credit card. Where countless apps help us monetize and access countless assets, from sofas, to spare rooms, to household appliances. It is "efficient" and "green" to optimize, we tell ourselves. We're sharing. Economists have terminology to codify the behaviors, attitudes and expectations of players faced with decisions of where to plug in, where to cash out, and how to game the system—terms like "rent seeking" and "social choice."

But not all of this sharing shares power, shares ownership, and allows for a reshaping of how we relate to the underlying property. Now is an important time to scratch off the scab on this topic, because there's a relationship between early "constitutional" decisions and the structures they create, and often the structures themselves make decisions. Disruption of hierarchy, patriarchy, corptocracy is clearly good—but disruption is not inherently a virtue.

All technology is not the same Lewis Mumford lights a path through the history of invention and social change with two useful words: polytechnic and monotechnic. Polytechnics derive from an ancient lineage and combine functions and skills shared by many people in a society. Polytechnic tools can be used in different ways and allow people to develop skills and exercise control over their work. In a polytechnic society no single method of doing anything dominates, and no authority dictates technology. Think of craftsmen with their various guilds employed to build a cathedral. Until the 17th Century, this polytechnic tradition performed the feat of transmitting the major technical heritage derived from the past, while introducing many fresh mechanical or chemical improvements, including inventions as radical as the spinning wheel and the printing press.

Monotechnics are defined by the concentration of authority in the proliferation of technology: "time-keeping, space-measuring, account-keeping, thus translating concrete objects and complex events into abstract qualities."[2] From there, these can be further transformed into commodities which can become derivatives, bundled and traded in haze of a financialize vapors.

Those structures created a monoculture. It is getting worse all the time, with some counties in the midwest planted over 90% with corn or soybeans, and that includes all other land uses—riparian, treeline, woodlots, pastures, lawns, roads, houses, parking lots and kidney dialysis centers. And we just keep paving. The Lowe's and Home Depot hang in hulking horizontal squander over the wetlands at the edge of town. Their over-fertilized, over-fungicided nursery plants leaking from the haphazard berms, dribbling out onto the landscape. The herons are not happy.

Of course much of our built environment in America agriculture, the barn, corral, pond and levy, was created to support the mixed family agriculture of the frontier and near-frontier. They used extremely high quality materials—glorious hand-hewn beams and old-growth redwood—for grape trellises, mangers and fenceposts. But these materials are grander, sometimes, than the paradigm defining their use or the insight of the original user. These are fenceposts worth re-tooling and re-thinking. They may be less permanent than the concrete-deco arches of an auto-metropolis like Detroit, they are gentler than the lasers and biotech we have now, but they are still tools of an export economy. Lest we romanticize those charming barns, we must remember how even these older forms drew from and degraded the commons.

For 60 years, the largest American export was cotton—a valuable chain of theft and cruelty, stimulated by greed and black tea. Grown by stolen Africans on stolen indigenous land, and fed to steam-powered factories in the sinking brick Empire of not-yet-unionized workers, many displaced from their own farms by the overproduction of exploited colonies. Indeed the cooperative efforts of share-croppers followed the logic of the Rochedale Group in Manchester, England, a group of factory workers collectively purchasing the strong-black-tea they need to sustain their working days. Agriculture built America, paid for the railroads and industrial revolution, amassed the wealth that was leveraged to drain the swamps, irrigate the deserts, and power the western portion of the continent behind mega-dams—it was enclosure, dispossession, exploitation that made it possible. It was the removal of resources from the common pool which filled the pockets of our nation and the wove the myths propping up

"Landshaft" describes as a working landscape, one that contains settlement, homesteads, wilderness, wood-lots, varieties of land-use in a harmonious, sensible configuration. This was the ideal of reformers in the 1840's, a kind of peaceful concession to seek balance between the wild, and domesticated worlds. In the words of Wendell Berry:

"The survival of wilderness—of places that we do not change, where we allow the existence even of creatures we perceive as dangerous—is necessary. Our sanity probably requires it. Whether we go to those places or not, we nee to know that they exist . And I would argue that we do not need just the great public wildernesses, but millions of small private or semi private ones. Every farm should have one; wildernesses can oc-cupy corners of factory grounds and city lots—places where nature is given a free hand, where no human work is done, where people go only as guests. These places function, I think, whether we intend them to or not, as sacred groves—places we respect and leave alone, not because we understand well what goes on there, but because we do not. "

—from *Home Economics*, Wendell Berry, North Point Press, 1987.

---

2 *Larding the Lean Earth*, Steven Stoll, Hill and Wang, 2002, 210

COMING MONEY TRUST

THE OCTOPUS—"ALDRICH PLAN"

the forefathers of our current capitalist superstructure.[3] That enclosure, which began with our natural resources, continues today with our networks, our data and our relationships.

Enclosure

James Scott, one of my favorite curmudgeons—and editor of American Georgics, the syllabus for the last Almanac—has put out another volume addressing some of the "format for freedom" questions that Thomas Paine has gotten me thinking about:

> "The question I want to pose is this: are the authoritarian and hierarchical characteristics of most contemporary life-world institutions—the family, the school, the factory, the office, the worksite—such that they produce a mild form of institutional neurosis? At one end of an institutional continuum one can place the total institutions that routinely destroy the autonomy and initiative of their subjects. At the other end of this continuum lies, perhaps, some ideal version of Jeffersonian democracy composed of independent, self-reliant, self-respecting, landowning farmers, managers of their own small enterprises, answerable to themselves, free of debt, and more generally with no institutional reason for servility or deference. Such free standing farmers, Jefferson thought, were the bases of a vigorous and independent public sphere where citizens could speak their mind without fear or favor. Somewhere in between these two poles likes the contemporary situation of most citizens of Western democracies: a relatively open public sphere but a quotidian institutional experience that is largely at cross purposes with the implicit assumptions behind this public sphere and encouraging and often rewarding caution, deference, servility, and conformity. …do the the cumulative effects of life within the patriarchal family, the state (GMAIL) and other hierarchical institutions produce a more passive subject who lacks the spontaneous capacity for mutuality so praised by both anarchist and liberal democratic theorists?
>
> If it does, then an urgent task of public policy is to foster institutions that expand the independence, autonomy, and capacities of the citizenry. How is it possible to adjust the institutional lifeworld of citizens so that it is more in keeping with the capacity for democratic citizenship?"
>
> —from *Two Cheers for Anarchy*, James Scott, Princeton University Press, 2012.

---

3 Rebecca Solnit has recently made some wonderful connections on this in an article in Harpers. http://www.harpers.org/archive/2014/08/the-octopus-and-its-children

Those blue bikes from the last section are sponsored and owned by a bank that gathers data about your commute on a computer attached to the public street. Boy are they convenient, but if the computer won't click open the lock, you're stuck. Are the owners at street-level? Are they available for a calm, spoken, rational negotiation about fairness? Cornel West says "justice is what love looks like in public," so what happens when decisions are remote from the public? Where administration of policies created by proxies is carried out by unthinking and unaccountable apparatus.

Last month, partially in reaction to the outcry in Ferguson Missouri, the Burlington, Vermont police department announced they would mount a video camera on the chest of their whole officer force. I'm reminded of the novel *The Circle*, by Dave Eggers, where a Google-Facebook-Apple corporation attempts to transform democracy via radical transparency of total digitization, total monitoring. In the half-wild west of today's internet, both protestors pay-to-play facebook posts bubble up and hype-merchants swarm for twitter rankings, anonymous hackers daylight Klu Klux Klan membership, credit cards are stolen, and the government keeps a carbon copy in their big twinkling bunker. It is not a gravitational force that drives us in this direction. Newton and Darwin would flinch at how we assign our minds to the "gravitational force" and "evolution" of the system in this direction.

The algorithms are artificial, even the ones running the voting machines. I've recently taken a survey of activists, friends of my mother and fairy godmothers—all of us have spam drama. Our messages don't get safely across the wild western internet lines. More and more activists' email accounts are hacked, calls are dropped, journalists working in Ferguson get their phones shut off, as we slowly realized the internet is actually telecom monopoly—a meta-technology of global capitalism—and we're too lazy to invest in alternatives.

This makes it easy to click around too much, to fall into a distracted frenzy. Today's techno innovation design cult (JFK/SFO) tend to hyperbolize their fantasies. The "internet of things" includes arduino-tended hydroponic vertical-glass-walled utopian skyscraper/arcologies. It's a narrative that insists we need "big data" to manage the vast land base that the technology's cost require. Its a new hyper-scale land-grab modality focused on pumping out a grid of predictable nutrition, lest the 9 billion rise up.

Who's idea of hopeful are rows of synthetic meat vats warmed by blinking servers, hovered over by drones of loving grace? Part Brave New World, part Pollyanna, if you tune into the media channels too much its hard to keep your lunch down. In highly abstracted high-rise condos, on expensive smelly gloss paper, these implausible scenarios, rendered with photoshop, are offered as a fashionista's parsley garnish—"modern farmer," "virtual grange," "sharecropping 2.0." Mirrored hallways of social media ring with rhetoric to "embrace the Anthropocene" and "if we're going to be Gods, lets get good at it."

Pinched between the Gates Foundation, National Geographic, and the massively-budgeted Farmers and Ranchers Alliance, "we must feed the world" merges with "the world must feed us." Yikes, it's a mafia land grab, an unconscious techno-feudalist, by Venture Capitalists for Venture Capitalists fantasia, but with good design and nice outdoor dinners. Slow Food for the 1%, and server-farm-incubated synthetic meat soylent green for everyone else.

You see how easy it is to fall down the dark hole of negativism. This is of course the reason we need an Almanac, to keep ourselves bound up in a fraternity of commitment—a positive outcome for all our concerted creative logistics for reasonable change. Greenhorns are dedicated to the project of holding space for producer (not consumer) culture—this volume is the sequel to our commitment in this direction. Manic, fruit-fly narratives of pop press give us little traction. I would argue that the Almanac format (dozens of contributors, thoughtfully compiled) is a far better context for literary discourse that can support our landscape-authorship, direct action and practical stewardship of our neighborhood ecosystems. It is a reestablishment of the information commons for which big tech continually gropes as they attempt to pull close our very exchanges and rope them off inside the proprietary platforms of the neo-capitalist sharing economy.

And what is the alternative to a tragic commons, to a broken fishery, a depleted pasture, a hazy inversion layer over the shared atmosphere of city after city after city? Clearly small is beautiful, and small towns are easier to police, but I want to explore the planning process it would take to "step down" into a feasible governance strategy for the thousands of decisions needed to shift out of the broken road trajectory.

Format of Transition

"Manage for what you want" says the charismatic genius Alan Savory. He's an ex-mercenary game park warden from Rhodesia who put forth the concept of "Holistic Management," a powerful decision-making framework for ranchers and land-managers working to optimize conditions of land-health, business-viability, and human happiness. It's a vector-approach to shifting the system, and a good way to visualize and measure out the "power down" from our current framework.

And we should not forget the other planes necessary for the trajectory either. Shakespeare had a more mythological concept map for the poison and its antidote:

> Sweet are the uses of adversity;
> Which, like the toad, ugly and venomous,
> Wears yet a precious jewel in his head.

—*As You Like It*, Act 2, Scene 1

The precious jewel is the antidote to the poison—in this case, the antidote to capitalism.

Who else is perched in the head of the poisonous toad of capitalism, dreaming a dream about a better world? Who are the precious, shameless, romantic, pragmatic, apocalyptic, optimistic jewels perched on the head of adversity? Young African Americans singing, crying, marching for justice against a militarized and unrestrained racist police-state. Young children of immigrants, migrants, and illegally-resident Americans stuck at the border, dislocated from their families, or stooping and and kneeling in the fungicide sponge of our California lettuce fields.

Essays in this volume shoot out like spider-webs across the wide chasm of impossibility—the glorious rescue mission whose daily requirements keep us human, hopeful and operating inside the scale of possibility.

What future can we realistically build together, will it need electricity? Will it need globalization? These practices we've found—biodynamics, permaculture, resilience breeding, state-change in the soil, reformats of ownership, reclaiming the value-chain, re-tooling for diversity, committing to lifetimes of partnership—how long before the allow us to reach steady-state? Do they require interns? Do they require servants making silicon chips? Do they require, absolutely require, the internet? Which technologies are relevant to, appropriate to, and gestating within the new agrarian mind? We've been provided with some answers and we've come up with some of our own.

So how might we execute them? What happens to a dream fulfilled? Turns out, it reeks of bat poop.

The Land

My dream hire, a young woman called Eliza Greenman, is a young orchardist without land and capital. She is a great organizer, hard worker, strong thinker, and a big-hearted sweetie pie. Trouble is, she's got a job offer to manage a bio-dome in the fortified desert palace of the prince of Abu-Dhabi. How can I, as measly director of a grassroots rat pack of agrarian cultural workers, ever compete with the prince of Abu Dhabi?

Let me wager an answer: free land.

What if I offered free land to the next generation of greenhorns as a legacy, or endowment? What if I gave Eliza the gift of good land, and not only to her, but to the the next greenhorn after her as well? What if I emancipated a piece of land from the commodity paradigm that currently rules our economy and planet.

Yes! A micro-retort against the macro-economy. An institutional elbow-jab. I've got the power for a tiny insurrection. Fifty-four thousand dollars is pretty good value for an 1820's house, 17 acres, 2 little barns, a glorious mountain view, and gravity-fed spring water. Owned by the bank and vacant for nearly three years, it's definitely a project house. The project is land reform.

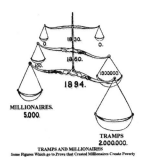

MILLIONAIRES.
5,000.

TRAMPS
2,000,000.

TRAMPS AND MILLIONAIRES
Some Figures Which go to Prove that Created Millionaires Create Poverty

Lets use this farm as a test-case, lets make these 17 acres and red brick farmhouse an emblem of everything. Lets make it a proxy-land for the total-land on which our whole economy is built. To do so is not such a stretch—all our human-scale and domestic micro-economies nest inside the macro-economy of the market.

We are programmed to see ourselves as "proprietarians" defined by what we control, what we obtain, what we own and oversee. We want to own land, and we understand innately the mathematics of territory + labor + taxes and maintenance = equity over time. But this is not the only way we operate as humans. Start with the family. There are the unpaid hours of our motherhood, the shared, non-monetized daily chores of house holding and reciprocity of the neighborhood. Think about functional societies in all cultures, and micro-societies that have non-economic relations of sharing and care taking and managing.

These are all powerful cultural paradigms, but in looking for an "authentic economy" on which to base my thinking I've found the context of our own economics to be more than befuddling. It usually comes back to exploitation, enclosure of the resources of the commons.

And the cream has certainly risen. Let me be specific. The Aluminum mining corporation Alcoa is worth 17.217 billion dollars.[4] It makes a material that we turn into disposable beverage containers. Alcoa owns assets of mountains which they dismantle for bauxite ore, as well as hydro-electric plants that generate the electricity needed to turn that land (destroyed via dynamite) into aluminum, which is useful for coke cans and has a use-value as a virgin material as well as a recycled material. It creates this value by adding energy, machines and applied chemistry to the power of mountains and rivers.

Meanwhile, there is a corporation called Twitter, valued at 27.3 billion dollars.[5] It operates on the internet, allowing users to post short messages which are visible to other users. These are called Tweets. Twitter has its own web-based software, but it relies on many external components of the information superhighway—hardware owned by server farms, internet service providers, home computer users, and telecommunications infrastructure of optic wires—all of which are made out of minerals and materials mined from the earth, also powered by electricity. Twitter does not charge for this service.

---

4 Alcoa's shares are trading at $14.64 (http://www.marketwatch.com/investing/stock/aa) and it has 1.176 billion shares outstanding (http://ycharts.com/companies/AA/shares_out-standing).
5 According to Reuters, as of July 30th of this year Twitter is worth about 27.3 billion dollars. (http://www.reuters.com/article/2014/07/30/us-twitter-results-idUSKBN-0FY27C20140730)

The value of these two companies is measured according to the vast internet of value assignments, which we call the market. Besides the Twitter and the Aluminum can, the fish in the sea and the salad in the field each have their own value assignment, which determines the prices of labor. These justify the invention and creation of machines, mortgages, and portable toilets for farmworkers. The terms of this economy dictate the "efficiencies of scale" for those parts of the economy which aren't considered valuable.

I require a tug

This is reflected on a micro level in a symptom which is more apparent for those of us examining our relationship to land and ownership—the end of the nest egg. Whereas in previous inter-generational interstices there was a little nudge, a dowry chest and maybe even a down payment, that's less than possible for increasing numbers of elderly people. No longer are they able to provide a little tug to get the next generation up to speed when they need every ounce of energy to bail out their own sinking financial boat.

Forced to liquidate, forced to extract and despoil, the market holds a moral grip on us. Its basic economics: what's under and inside the mountain (gasoline, bauxite, shale-gas) is more valuable than what grows on top (acorns, wild boar) and can be gently, sustainably harvested. So we blow up the mountains, make the cans and fill them with corn syrup from corn grown in some other, far cheaper field. This is the logic we live under, and it leads to bankruptcy.

Because at a certain moment, this economy ends. And I think it ends first with Agriculture. Agriculture, the intersection of economy and ecology, between culture and industry, is the right place for the inter-generational truce that is necessary for restoration—which started out the trajectory of this essay.

At a certain moment the land becomes irreplaceable as the place we need to grow our food. We are reaching that point, and a hard stare at the state of the world proves to us that our prime farmland—especially the land close to centers of population, well watered by natural rainfall, and decently healthy—must be exempted from the global shuffle of trade-offs. We need a new economics that values it for what it can provide over the long run, And the height of that value—the highest, best use—is for good, clean, food.

The land is true wealth because the land holds our whole society and culture, our whole ecology. I frame my little piece of land in these sweeping terms because so often land is traded and grabbed as an abstraction, and to talk about this tiny patch in the context of its grandiosity is a kind of poetic justice. This total-land has value in our economy, and the thought experiment we are engaged in here is about how that value is created, assigned and managed, and how we can organize our systems of human culture and economy to suit the best interests of the land on which they rest.

Which brings us back to my own little land. It just so happens that the town where my particular land retort is located—Westport, NY—is about contemporary with

Thomas Paine, whose treatise *Agrarian Justice* provided a basis for much subsequent land reform discussion. Paine released this treatise as a retort to a sermon given by the influential Bishop Landalf. In that sermon, reflecting the predominant owning-class perspective of his parishioners, Landalf cited the "Wisdom of God in having made both rich and poor," and endorsed social inequity as part of the natural order of the universe.

The church has a long history of noticing that the meek will inherit the earth, but in that moment the Church was weighing in on the side of the mighty. In another time of both extreme poverty and wealth, Landalf was conferring upon this situation the wisdom of God—validating it in a rational context, as if this social condition were an outcome of a natural law. The laws of gravity (discovered by Newton) and later the laws of evolution (discovered by Darwin) of the emerging scientific age had come to dominate the rhetoric and worldview of leaders. Paine felt it was critical that gross injustice was not assumed as a baseline context. Poverty and destitution, he reminded, are not a natural phenomena like gravity and orbits, evolution and diffusion. This assertion remains relevant today, with our levels of inequality even further pronounced than at that time.

Paine would point out that we cannot create land and we cannot destroy it, and that ownership is really just the "right to use" the land. Our current economy is a kind of suspended vinaigrette of small, medium and large versions of businesses, landowners and national actors. But it is a vinaigrette out of balance. It doesn't hold together, it doesn't match the reality of the planet. It's a clumpy, lumpy mess. To remedy this, Paine proposed an equal payment system in which citizens would be repaid yearly for the inheritance—the land—which they had lost. Henry George echoed this idea with his own proposal for a single tax on land value, intended to reflect the fact that the economic value of that land and its resources is a commons, which belongs equally to all people. And I'm proposing to restructure our models of land ownership and use with a re-fortification of the commons, to shore up the communities and the lifeways which depend on this land.

A Community Format

This little red house, and the land it sits on, is a chance to experiment with radical de-coupling of finance and real estate, and a way to reframe "financial wealth" into "community wealth." Positioned between the train tracks and Lake Champlain, our little town has a county fair ground, a victorian bandstand,[6] two marinas, a hiking trail trust,[7] and organic grain mill,[8] and a little grocery store that daily bakes their own hamburger buns. It is the romantic all-American town built to serve the extraction economy of lumber, iron, and stone on territory of the Iroquois and Western Abenaki.

6   http://www.ballardparkny.org/events.html
7   http://www.champlainareatrails.com/
8   http://champlainvalleymilling.com/

This is the site of our 17 acres. It may be a small gift, but its a land gift.

By making a "Land Gift" I can give my little mission-driven non profit organization (and the young farmers movement it serves) a lasting legacy of free land. Afterwards (since perpetuity is a long time) the land will become a part of the larger Adirondack Community Housing Trust.[9] This means finally! We'll have a stable place for our library of 8,000 books, our tool-shop, props and event-production materials. As long as Greenhorns exists, it will have a home. When, and if, it ceases to exist, the house will become permanently affordable housing for farmers new to the region.

The Community Land Trust model was designed by Bob Swan—inspired by the teachings of E.F. Schumacher—while he was working at Indian Line Farm and New Communities, a 5,700 acre cooperative farm for landless African American farmers on a former slave plantation in Georgia created by the International Independence Institute and the Sherrod family.[10] At the time, the latter was one of the largest black-owned tracts of farmland in the U.S., but it was was eventually lost due to discrimination on the part of USDA lenders. The loss triggered a major mobilization[11] to save black-owned farms, centered in North Carolina. It wasn't until 20 years later, when the Pigford v. Glickman case was settled, that the Sherrods received compensation, allowing them to embark on the next chapter, and purchase a different plantation, now called Resora.[12]

More and more of the young farmer community is studying up on land reform. Why? Because hardly anyone can pay for the land they farm on by farming. Hardly anyone can find a deal as good as I did in America today. In order for our movement to succeed (and provide food sovereignty in the places we farm) a thousand times more farmers are going to need the confidence, capital, and soil-health that you get from secure tenure. In America, this usually means ownership. Mark Twain famously quipped "buy land, they aren't making it anymore."

And though ownership seems the only sure-thing-paradigm for the majority of Americans, there's a pinch—we mostly cannot reach it due to student loan debt, low pay, inadequate savings and the tricky dilemma of the impossibility of capitalizing a business, a family and a land-mortgage at the same time. In an ideal world your parents would have spent 20 years building the business, barns, brand and land-health and handed it to you, debt free. The pinch that brings us back to reality, is that new entrants to agriculture don't have that cushion, and neither does society. The pinch for the farmer is a pinch for society in general.

---

9   http://www.adkhousing.org/
10 http://en.wikipedia.org/wiki/Robert_Swann_(land_trust_pioneer)
11 http://landloss.org/
12 http://www.wkkf.org/what-we-do/featured-work/the-arc-of-justice-bends-toward-cypress-pond

There was a famous study in the 1940's performed by USDA agricultural economist Walter Goldschmidt. It was commissioned because the USDA was considering constraining the size of operations who would be eligible for subsidized irrigation. He looked at 2 towns in the Central valley of California, Arvinda and Dinuba. Both towns produced vegetables of similar types, volumes, values, and sold to similar markets. The difference was ownership of land from town to town: in one, majority owner-operated, in the other majority absentee-owned. You may be able to imagine the outcome. In terms of education, civic institutions, road qualities, health outcomes, and church attendance the owner-operated town outranked the absentee-town in every category. The USDA suppressed the study, defunded the agency which created it, and never investigated land tenure again.

Goldschmidt interpreted his finding this way: "it is difficult to prove causation in history, for each society is unique and the forces are complex, but there are few who doubt that the nature of rural land tenure is intimately related to the character of the social order."[1]

At stake was a US policy that worked to distribute the wealth created by tax-dollar-funded irrigation projects. Seemingly the policy recognized that the state, as an instrument of democracy, has the responsibility to distribute the benefits of expenditures democratically. Unfortunately, this was not the outcome and California descended into a cronyism we today take for granted.

---

1   *The People's Land*, Barnes, Rodale 1975, 171.

My land is OUR Land. And as discussed at the outset of this essay, we face an unprecedented risk to our national security when it comes to land ownership. An estimated 400 million acres will change hands in the next 20 years. That's an opportunity, to change the management of that land from commodity-production for national or international markets, to sustainable, careful, caring agriculture that produces food for local or regional markets. But it's also a danger, because if no one shows up to love this land, with a plan of how to capitalize and run a business on it "in the real world," it will become a commodity. The reason is human—farmers are aging. More than 70 percent of American farmland is owned by those 65 and older.

What comes first, the detailed chicken, or the deviled eggs? The history and frustrating economics of agriculture—land too expensive, labor too cheap, energy costs and subsidies distorted—could take up the rest of this essay. Farmers who've survived the perverse unreality which is our mainstream American food system have had to invent functional structures. To be more succinct, lets call this a "format." The local food movement, and local food explosion, is made possible by the "format" of very specific business structures and strategies that meet the needs of both consumers and producers. Lets focus on some familiar ones: Farmers markets and C.S.A.s (Community Supported Agriculture—where the customers pre-pay for their produce and get one box per week). You probably know this story, farmers are able to capture the full retail dollar, charge reasonable prices for their product based on quality, relationship, and ecological and values-based stewardship. Similarly, with C.S.A., the economic mechanism, or "format," puts cash into the farmer's hand at the beginning of the season so that they don't have to go into debt, and distributes the risks of agriculture to the whole community. It also benefits the consumers who get a good and reliable deal on produce.

My little red house in this fairy tale is a relic of a previous economy—the extractive boom-days of mining, lumbering, and quarrying in the Adirondacks. That economy is now imploded, making this land and house accessible and affordable for redemption by sustainable agriculture. So how can I prevent it from ever entering the boom again, and becoming a slot for a shale gas warehouse? What do we learn from this study of the "format" of C.S.A. and Farmers Market, that we can apply to the troublesome quandary of land access for new growers?

Agrarian Trust

When I went to visit Wendell Berry, I arrived with an arm load of questions for him, centered on this fundamental dilemma about land access. He chuckled at me and quipped, "Young lady, there is no big solution, only many thousands of small ones." Looking forward at the thousands of small transactional moments that will decide the destiny of our collective farmland, it feels like an inverted riddle from the old agrarian sphinx. But one worth cracking. One nut at a time. It will take a whole team of hopeful squirrels.

There will be many forms to this solution, all of them at their core a kind of truce between the generations, expressed both economically and territorially. Donella Meadows counsels us to "expose [our] mental models to open air." As Americans that also sometimes means opening up to models developed in other countries. The C.S.A. for instance, was first practiced in Japan, then Germany, and only arrived in the U.S.A. in the early 1980's via the international biodynamic farming network. So in looking for a template for a solution, we were lucky to find a project in France with the same set of principles. It is called Terre De Liens,[13] and it already holds more than 100 farms, on more than 6,000 ha with a value of over 53 million euros. The average investment size is 5,000 euros—some of the farmers gifted the land outright, some of them received "shares" equivalent to the value of their land which they can pass to their children, who can slowly liquidate the equity over time. This is a perfect locus for a conversation about ecology, economics, and resource management on our good planet—an "economic truce" between the generations.

The truce I'm committed to help build is called Agrarian Trust. This will be a trust that can receive donations of land and money with which to purchase land. It will put a covenant on the deed of that land that it will be protected for sustainable farming, in perpetuity. It can only ever be farmed for local sale. The idea is to give farmers a lifetime lease on the land, providing the benefits and security of ownership, without the cost. The point of this work is to demonstrate that keeping sustainable farmland in farming, is highest purpose for that land, and for the community that surrounds it. The Farmland Commons we are building—with help from the Berkeley, California-based Sustainable Economies Law Center[14]—will be the legal container for a land reform. For Agrarian Trust to succeed, we need thousands of grandmothers, landowners, and benevolent investors to give land and equity to the trust, and to pass forward a legacy of opportunity for new farmers. This will require a major shift in how we perceive land, and treat it, both culturally and economically, as a commons which we share with the future.

In studying this issue and providing immediate support for the stakeholders, Agrarian Trust has built a library of resources for land transfer,[15] and a community of farm-service providers committed to systemic change. This means stories of access, strategies of equity-sharing employed by entering and exiting farmers and their partners on the land. It means tracking the personal trajectories and negotiations that reject the logic of the macro-economy and accommodating the needs of both parties—those entering and those exiting. Again, these are humans we are talking about, humans who were raised by unpaid mothers and who take care of their own children for free. As Joel Salatin puts it "we can't get out of farming, until y'all get in." In case after case, we see motivation on the retiring side that goes beyond retirement funds and "doing right by

" .. there remained, firmly embedded in the very act that gave birth to the Bureau of Reclamation, that explicit clause designed to spread the benefits of public subsidies among the maximum number of resident farmers. Few might have noticed but for an economist at the University of California whom one historian has called " the last Jeffersonian". Paul Taylor was shocked by the disparity of wealth he found in California, which he, like Henry George, traced to land monopoly. In a series of elegantly concise legal briefs and articles, Professor Taylor fought a long and often lonely battle to make the government enforce its own law."

—from *Farewell, Promised Land*, Dawson and Brechin, University of California, 1999, 161.

400 MILLION ACRES

LAND IN TRANSITION

---

13   http://www.terredeliens.org/
14   http://www.theselc.org/
15   http://www.agrariantrust.org/resources

the children." There is a willingness to experiment with arrangements that consider the land's best interests as well.

People who are disconnected from land by pavement, elevators, and swivel-chair notions sometimes forget about the specific, connected power of the land on which they are working—the land that feeds them, the land that feeds the oxygen they breathe. Agrarian Trust is one approach, built on the lineage and analysis of Tolstoy, Paine, Barnes, Swan, the Levellers. Many dozens more methodologies may be created for passing land forward—a range of partnerships, community investment models, logical, family-like approaches.

According to the recent U.S. Department of Agriculture census, the vast majority of farmers—both on commodity-producing and community-based farms—are subsidized by other form of income: a wife's teaching job, a government pension, a marijuana crop, or side business plowing snow or selling tractors. Farm businesses that have the option often keep themselves afloat and capitalized by selling lots for housing development.

The context in which many food producers are running their businesses tends to be colored, more than we'd sometimes like to admit, by the global economy, the commodity market, and market price. These are all factors which impact the discrete business decisions that are being made on the farm. The big picture shows us that natural resources, human labor, and agriculture in general is profoundly undervalued, particularly in comparison to other sectors (like technology) which command a far better set of margins despite their intangible nature.

We know that farmers shouldn't be economically undermined by real estate and development pressures that over-ride their stewardship practices. Therefore, we need to take away the value of the land for the individual, and put the value into the hands of the community (OUR land) that benefits from the food, a bit like a C.S.A. In addition we must remove the speculative forces from that land (which would consider it for its recreational, scenic, development or other non-agricultural purposes). We recognize that for good farmland, good farming is its perfect destiny, and should be its permanent use.

To my ear, the thousands of small solutions Berry spoke of means thousands of small farms and local economies, each adapted to the farmsteads, landscapes, and cultures of this large land.

At the Greenhorns, we've worked to connect young farmers while providing social support and cultural enlivenment. Now, we also hope to help them find land. That's why we've founded Agrarian Trust, an organization dedicated to improving land access for sustainable farmers, and to helping those who hold the land pass it forward graciously, to new stewards who will grow food for their local economies.

Onward

These thousands of small solutions, small farms, small economies and small communities may not make the whole "step-down" by themselves—there will certainly be more required, and the effort must be sustained. But there is value in micro-retorts against macro-economies and their foundations, their histories of enclosure and exploitation. There is value—both symbolic and tangible, economic value—in re-establishment of the commons and a defense of its borders. We must be flexible, these borders will change as anything organic does. They may recede, they may shift, but the quiet momentum, the diversity, of thousands of small solutions is a model with far deeper roots than the strange monoculture that so many in the industrialized world have come to know. And as we hack these new paths through to allies on the other side of the hedge, we also discover paths long-worn, paths we had sometimes missed on the first pass along the edge of the field. So functions the Almanac in front of you: guide yourself to something new, learn something long-established. And keep moving forward.

Index of Contributors

Josef Beery
Printmaker
Free Union, VA
www.josefbeery.com

Janna Berger Siller
Organic farmer
Falls Village, CT

Audrey Berman
Farmer and organizer
Hudson, NY

Kayle Brandon
Artist
Bristol, UK
www.cube-cola.org

Jayme Bregman
Ebullient American expatriate
Kerepehi, New Zealand

Tess Brown-Lavoie
Farmer and writer
Providence, RI
www.sidewalkendsfarm.wordpress.com

Elizabeth Brownlee
Farmer and naturalist
Crothersville, IN
www.nightfallfarm.com

Nate Chisholm
Livestock and ecosystem manager
Glen Ellen, CA

Marada Cook
Food systems entrepreneur
Vasselboro, ME
www.crownofmainecoop.com

Jacob Cowgill
Vegetable and grain farmer
Power, MT
www.PrairieHeritageFarm.com

Savitri D
Anti-consumerist gospel shouter
New York, NY

Douglass DeCandia
Grower and food program coordinator
Waccabuc, NY
www.foodbankforwestchester.org

Anita Deming
Executive Director,
Cornell Cooperative Extension
Essex County, NY
www.adirondackharvest.com
www.ccenny.com

Brian Dewan
Multi-media artist
Catskill, NY
www.dewanatron.com

William Duesing
Farmer, author, activist
Oxford, CT
www.ctnofa1982.blogspot.com

Jenny Edwards
Aesthetic explorer
Berkeley, CA

Lucas Foglia
Photographer
Berkeley, CA
www.lucasfoglia.com

Amy Franceschini
Artist and Futurefarmer
San Francisco, CA
www.futurefarmers.com

William Giordano
Farmer, composer, designer, musician
Rockport, ME
www.tilth.me

Max Godfrey
Traditional singer, farm apprentice
Atlanta, GA
www.facebook.com/maxgodfreymusic

Aimee Good
Artist, Educator, Farmer
Brooklyn, NY
www.gooddirtgarlic.com

Jen Griffith
Farmer, boat-lover, banjo struggler
Glen Oaks, NY

Jeff Hake
Farm training program manager
Champaign, IL
www.farmersfold.tumblr.com

Katelyn Hale
Artist, educator, farmer
Portland, OR
www.curiocityprinting.tumblr.com

Shannon Hayes
Farmer, author
Richmondville, NY
www.TheRadicalHomemaker.net

Elizabeth Henderson
Organic farmer, writer, ag-tivist
Newark, NY
www.peaceworkcsa.org
www.nofany.org

Heidi Herrmann
Farm and teal truck owner
Healdsburg, CA
www.strongarmfarm.com

Dougald Hine
Dark Mountain co-founder
Västerås, Sweden
www.dark-mountain.net

Index of Contributors

Fritz Horstman
Art, music, science!
Bethany, CT
www.fritzhorstman.com

Rebecca Gayle Howell
Author, poetry editor
Provincetown, MA
www.rebeccagaylehowell.com

Adam Huggins
Phytocurious impermaculturalist
Diablo Valley, Ohlone Territory
www.sunfishmoonlight.wordpress.com

Qayyum Johnson
Buddhist Vegetable Farmer
Green Gulch Farm, CA
www.donotdelay.wordpress.com

Rose Karabush
Vegetable farmer
Amenia, NY
www.maitrifarmny.com

Patrick Kiley
Publisher and writer
Hudson, NY
www.publicationstudio.biz

C. R. Lawn
Fedco Seeds Coop founder
Waterville, ME
www.fedcoseeds.com

Bobby Losh-Jones
Farmer, baker
Gordon, GA
www.babeandsagefarm.com

Alex MacLean
Aerial photographer
Lincoln, MA
www.alexmaclean.com

Jonathan Magee
Farmer and community organizer
Western Massachusetts

Ginny Maki
Artist
Portland, OR
www.ginnymaki.com

Willa Mamet
Photographer
Oakland, CA
www.willamametphotography.com

Julia McDonald
CSA farmer
Manteno, IL
www.peasantsplot.com

Colin McMullan
Odd jobs
Storrs, CT
www.emceecm.com

Katharine Millonzi
Ethnobotanist, writer, and gastronome
Hudson, NY

James Most
Chestnut Promotion and Propagation
Orcas Island, WA

Leaf Myczack
Teacher of earth stewardship
Blue Ridge Mountains, VA
www.sustainability-teaching-farm.com

Schirin Rachel Oeding
Farmer and handcrafter
Germany & Denmark

Toni Ortega
Farmer, Lost City co-founder
Reno, NV
www.lostcityfarm.com

Ali Palm
Restoration horticulturalist + goatherder
Maine

Byron Palmer
Grassland manager
Penngrove, CA
www.sonomamountaininstitute.org

Jesse D. Palmer (PB Floyd)
Author, activist, attorney
Berkeley, CA
www.slingshot.tao.ca

Alison Parker
Organic farmer + herbalist
Libertyville, IL
www.radicalrootfarm.com

Pat Perry
Artist
Detroit, MI
www.patperry.net

Megan Prelinger
Public historian,
library-builder, naturalist
San Francisco, CA
www.prelingerlibrary.org

Rick Prelinger
Archivist, filmmaker, professor
San Francisco, CA
www.prelingerlibrary.org

Wesley Price
Mycologist
Cape Cod, MA
www.efungi.com

Ernesto Pujol
Performance artist and
social choreographer
Old Chatham, NY
www.ernestopujol.org

Douglas Rainsford Tompkins
www.tompkinsconservation.org

Kate Rich
Artist + Trader
Bristol, UK
www.cube-cola.org

Eli Rogosa
Biodiversity farmer, sailor
Colrain, MA
www.growseed.org

Alanna Rose
Artist and farmer
Plainfield, NY
www.alannarose.com
www.cairncrestfarm.com

Lori Rotenberk
Environment + sustainability journalist
Stevanston, IL
@loriandwhillie

Marice Sariola
Iconographer
Dunsborough, Australia
www.iconsbymarice.com.au

Kat Shiffler
Agroecologist and beekeper
Lincoln, NE
www.sweetberthas.com

Alexandra Sing
Photographer

Darby Smith
Biodynamic Farmer
Blairsville, GA
www.thesundogfarm.com

Erin Smith
Farm logo + label designer
Portland, ME
www.wheatberrydesign.com

Steve Sprinkel
Farmer and Cook
Ojai, CA
www.farmerandcook.com

Connor Stedman
Wilderness educator +
ecological designer
Hudson River Valley, NY
www.renewingthecommons.wordpress.com

Sonja Swift
Freelance writer
San Francisco, CA

Ariana Taylor-Stanley
farmer and "ag-tivist"
Seattle, WA
www.arianataylorstanley.com

Pete Tridish
Philadelphia, PA
www.imarad.io

Natsuko Uchino
Artist
Paris, France
www.greenteagalleryworldwide.com
www.natsuko2point0.tumblr.com

Rachel Weaver
Young farmer + philosopher
Denton, TX

Jean Willoughby
Project coordinator at RAFI
Pittsboro, NC

Joseph Wolyniak
DPhil candidate, scholar
Denver, CO

Alexis Zimba-Kirby
Farmer, Baker, Chef
Midcoast Maine

Helen Zuman
Brooklyn, NY

INTRODUCTORY NOTE

The astronomical data in this booklet are expressed in the scale of universal time (UT); this is also known as Greenwich mean time (GMT) and is the standard time of the Greenwich meridian (0° of longitude). A time in UT may be converted to local mean time by the addition of east longitude (or subtraction of west longitude), where the longitude of the place is expressed in time-measure at the rate of 1 hour for every 15°. The differences between standard times and UT are indicated in the chart on page 22; local clock times may, however, differ from these standard times, especially in summer when clocks are often advanced by 1 hour.

PRINCIPAL PHENOMENA OF SUN AND MOON, 2015

THE SUN

		d h			d h m		d h m
Perigee	... Jan.	4 07	Equinoxes	... Mar.	20 22 45 ...	... Sept.	23 08 21
Apogee	... July	6 20	Solstices	... June	21 16 38 ...	... Dec.	22 04 48

PHASES OF THE MOON

Lunation	New Moon			First Quarter			Full Moon			Last Quarter		
		d	h m		d	h m		d	h m		d	h m
1138							Jan.	5	04 53	Jan.	13	09 46
1139	Jan.	20	13 14	Jan.	27	04 48	Feb.	3	23 09	Feb.	12	03 50
1140	Feb.	18	23 47	Feb.	25	17 14	Mar.	5	18 05	Mar.	13	17 48
1141	Mar.	20	09 36	Mar.	27	07 43	Apr.	4	12 06	Apr.	12	03 44
1142	Apr.	18	18 57	Apr.	25	23 55	May	4	03 42	May	11	10 36
1143	May	18	04 13	May	25	17 19	June	2	16 19	June	9	15 42
1144	June	16	14 05	June	24	11 03	July	2	02 20	July	8	20 24
1145	July	16	01 24	July	24	04 04	July	31	10 43	Aug.	7	02 03
1146	Aug.	14	14 53	Aug.	22	19 31	Aug.	29	18 35	Sept.	5	09 54
1147	Sept.	13	06 41	Sept.	21	08 59	Sept.	28	02 50	Oct.	4	21 06
1148	Oct.	13	00 06	Oct.	20	20 31	Oct.	27	12 05	Nov.	3	12 24
1149	Nov.	11	17 47	Nov.	19	06 27	Nov.	25	22 44	Dec.	3	07 40
1150	Dec.	11	10 29	Dec.	18	15 14	Dec.	25	11 11			

ECLIPSES

A total eclipse of the Sun	Mar. 20	Greenland, Iceland, Europe, North Africa and north-western Asia.
A total eclipse of the Moon	Apr. 4	The western half of North America, Oceania, Australasia and eastern Asia.
A partial eclipse of the Sun	Sept. 13	Parts of southern Africa, the southern half of Madagascar, the southern Indian Ocean and the eastern part of Antarctica.
A total eclipse of the Moon	Sept. 28	Western Asia, Africa, Europe, the Americas excluding the western half of Alaska.

LUNAR PHENOMENA, 2015

MOON AT PERIGEE

	d h		d h		d h
Jan.	21 20	June	10 05	Oct.	26 13
Feb.	19 07	July	5 19	Nov.	23 20
Mar.	19 20	Aug.	2 10	Dec.	21 09
Apr.	17 04	Aug.	30 15		
May	15 00	Sept.	28 02		

MOON AT APOGEE

	d h		d h		d h
Jan.	9 18	May	26 22	Oct.	11 13
Feb.	6 06	June	23 17	Nov.	7 22
Mar.	5 08	July	21 11	Dec.	5 15
Apr.	1 13	Aug.	18 03		
Apr.	29 04	Sept.	14 11		

OCCULTATIONS OF PLANETS AND BRIGHT STARS BY THE MOON

Date		Body	Areas of Visibility
	d h		
Jan. 25	12	Uranus	N. half of Africa, S. Europe, Middle East, Russia, N. Asia
Jan. 29	18	*Aldebaran*	Northernmost Canada
Feb. 21	22	Uranus	North Polynesia, USA except north west, Mexico
Feb. 25	23	*Aldebaran*	Alaska, N.W. Canada, northernmost Russia, Greenland, Iceland, Scandinavia
Mar. 21	11	Uranus	Easternmost Brazil, Central Africa, Middle East, W. Asia
Mar. 21	22	Mars	Parts of Western Antarctica, S.W. South America
Mar. 25	7	*Aldebaran*	Kazakhstan, Russia, N.E. Scandinavia, extreme N.E. China, N. Greenland, N.W. Canada, Alaska
Apr. 21	17	*Aldebaran*	N.W. USA, Canada, Greenland, Iceland, Scandinavia, extreme N. of British Isles, N.W. Russia
Apr. 26	7	Juno	Eastern S.E. Asia, N. Papua New Guinea, Micronesia, N. Melanesia, French Polynesia
May 15	12	Uranus	Central South America, West and Central Africa
June 11	20	Uranus	S. and E. Australia, New Zealand, Fiji, Samoa, French Polynesia
June 15	2	Mercury	S. tip of India, Sri Lanka, most of S.E. Asia, Micronesia
June 15	12	*Aldebaran*	East and North Canada, Greenland, Iceland, N. Scandinavia, North and central Russia
July 9	3	Uranus	E. parts of Antarctica, S. Indian Ocean, S. tip of Madagascar, westernmost Australia
July 12	18	*Aldebaran*	N. Japan, E. Russia, Alaska, N. Canada, Greenland, Iceland
July 19	1	Venus	New Guinea, N.E. Australia, Melanesia, French Polynesia
Aug. 5	9	Uranus	Antarctic Peninsula, S. South America, Falkland Islands
Aug. 9	0	*Aldebaran*	Middle East, E. Europe, N.W. Asia, Scandinavia, Russia, Alaska, N.W. Canada
Sep. 1	16	Uranus	Wilkes Land, Victoria Land, most of New Zealand
Sep. 5	6	*Aldebaran*	Eastern North America, Europe, western Russia, N.W. Asia
Sep. 29	1	Uranus	Parts of Antarctica, South Africa, S. tip of Madagascar
Oct. 2	13	*Aldebaran*	Micronesia, Japan, North America
Oct. 8	21	Venus	Australia, E. Melanesia, New Zealand, Victoria Land
Oct. 11	12	Mercury	S. South America, Falkland Islands, parts of Antarctica
Oct. 26	10	Uranus	E. part of Antarctica, New Zealand, S. French Polynesia
Oct. 29	23	*Aldebaran*	N.W. Africa, Europe, Russia, N. Middle East, N. Asia
Nov. 22	19	Uranus	Queen Maud Land, Enderby Land, southern Indian Ocean
Nov. 26	10	*Aldebaran*	Japan, Eastern Russia, N. USA, Canada, Greenland
Dec. 6	3	Mars	Central and East Africa, S. Arabian Peninsula, S. tip of India, Indonesia, Australia
Dec. 7	17	Venus	North and Central America, Caribbean
Dec. 20	1	Uranus	Antarctic Peninsula, S. tip of South America, Falkland Islands
Dec. 23	20	*Aldebaran*	Easternmost coast of Canada, N.W. Africa, Europe, Russia, northern Asia

PLANETARY PHENOMENA, 2015

GEOCENTRIC PHENOMENA

MERCURY

	d h	d h	d h	d h
Greatest elongation East	Jan. 14 20 (19°)	May 7 05 (21°)	Sept. 4 10 (27°)	Dec. 29 03 (20°)
Stationary	Jan. 21 04	May 19 11	Sept. 17 13	—
Inferior conjunction ...	Jan. 30 14	May 30 17	Sept. 30 15	—
Stationary	Feb. 11 07	June 11 20	Oct. 8 22	—
Greatest elongation West	Feb. 24 16 (27°)	June 24 17 (22°)	Oct. 16 03 (18°)	—
Superior conjunction ...	Apr. 10 04	July 23 19	Nov. 17 15	—

VENUS

	d h		d h
Greatest elongation East	June 6 18 (45°)	Stationary	Sept. 5 09
Greatest illuminated extent	July 10 04	Greatest illuminated extent	Sept.21 15
Stationary	July 23 06	Greatest elongation West	Oct. 26 07 (46°)
Inferior conjunction ...	Aug. 15 19		

EARTH

	d h		d h m		d h m
Perihelion ... Jan. 4 07		Equinoxes ... Mar. 20 22 45 ...		... Sept. 23 08 21	
Aphelion ... July 6 20		Solstices ... June 21 16 38 ...		... Dec. 22 04 48	

SUPERIOR PLANETS

	Conjunction	Stationary	Opposition	Stationary
	d h	d h	d h	d h
Mars	June 14 16	—	—	—
Jupiter	Aug. 26 22	—	Feb. 6 18	Apr. 8 20
Saturn	Nov. 30 00	Mar. 14 22	May 23 02	Aug. 2 20
Uranus	Apr. 6 14	July 26 16	Oct. 12 04	Dec. 26 11
Neptune	Feb. 26 05	June 12 20	Sept. 1 04	Nov. 18 21

The vertical bars indicate where the dates for the planet are not in chronological order.

HELIOCENTRIC PHENOMENA

	Perihelion	Aphelion	Ascending Node	Greatest Lat. North	Descending Node	Greatest Lat. South
Mercury	Jan. 21	Mar. 6	Jan. 17	Feb. 1	Feb. 24	Mar. 27
	Apr. 19	June 2	Apr. 15	Apr. 30	May 23	June 23
	July 16	Aug. 29	July 12	July 27	Aug. 19	Sept. 19
	Oct. 12	Nov. 25	Oct. 8	Oct. 22	Nov. 15	Dec. 15
Venus	Apr. 18	Aug. 8	Mar. 15	May 10	July 5	Jan. 18
	Nov. 29	—	Oct. 26	Dec. 20	—	Aug. 31
Mars	—	Nov. 20	Apr. 12	Oct. 13	—	—

Jupiter, Saturn, Uranus, Neptune: None in 2015

Predicted Perihelion Passages Of Comets, 2015

	T	q(au)	P(y)		T	q(au)	P(y)
141P/Machholz	Aug. 24	0·76	5·25	10P/Tempel	Nov. 14	1·42	5·36
61P/Shajn-Schaldach	Oct. 2	2·11	7·06	230P/LINEAR	Nov. 18	1·49	6·27
151P/Helin	Oct. 8	2·47	13·90	249P/LINEAR	Nov. 26	0·50	4·59
P/2001 H5 (NEAT)	Oct. 21	2·44	15·04	P/2010 R2 (La Sagra)	Nov. 30	2·62	5·45
P/2007 V2 (Hill)	Oct. 23	2·78	8·22	P/2003 WC7	Dec. 5	1·66	11·79
P/1994 N2 (McNaught-Hartley)	Oct. 24	2·45	20·61	(LINEAR-Catalina)			
22P/Kopff	Oct. 25	1·56	6·40	P/2002 Q1 (Van Ness)	Dec. 10	1·56	6·73
P/2005 RV25	Oct. 28	3·58	8·94	204P/LINEAR-NEAT	Dec. 11	1·93	6·99
(LONEOS-Christensen)				180P/NEAT	Dec. 12	2·49	7·58
P/2008 Y2 (Gibbs)	Nov. 5	1·63	6·78	P/1998 QP54	Dec. 25	1·89	8·62
214P/LINEAR	Nov. 12	1·85	6·87	(LONEOS-Tucker)			

PHENOMENA, 2015

VISIBILITY OF PLANETS

MERCURY can only be seen low in the east before sunrise, or low in the west after sunset (about the time of beginning or end of civil twilight). It is visible in the mornings between the following approximate dates: February 6 to April 1, June 9 to July 16 and October 7 to November 3. The planet is brighter at the end of each period,(the best conditions in northern latitudes occur in mid-October and in southern latitudes from mid-February to mid-March). It is visible in the evenings between the following approximate dates: January 1 to January 24, April 18 to May 21, August 1 to September 24 and December 5 to December 31. The planet is brighter at the beginning of each period,(the best conditions in northern latitudes occur from late April to early May and in southern latitudes from mid-August to mid-September).

VENUS is a brilliant object in the evening sky from the beginning of the year until in the second week of August it becomes too close to the Sun for observation. From the end of the third week of August it reappears in the morning sky where it stays until the end of the year. Venus is in conjunction with Mercury on August 5, with Mars on February 21, August 29 and November 3 and with Jupiter on July 1, July 31 and October 26.

MARS is visible as a reddish object in Capricornus in the evening sky at the beginning of the year. Its eastward elongation gradually decreases as it moves through Aquarius from early January, Pisces from mid-February, briefly into Cetus in early March, then into Pisces again and into Aries in late March. It becomes too close to the Sun for observation in mid-April. It reappears in the morning sky during the first week of August in Gemini and then moves into Cancer in early August, Leo from early September (passing 0°.8 N of *Regulus* on September 24) and into Virgo early in November, where it remains for the rest of the year (passing 4° N of *Spica* on December 21). Mars is in conjunction with Venus on February 21, August 29 and November 3 and with Jupiter on October 17.

JUPITER can be seen for most of the night in Leo at the beginning of the year. Its westward elongation gradually increases, passes into Cancer in early February, and is at opposition on February 6 when it can be seen throughout the night. Its eastward elongation then gradually decreases and from mid-May it can be seen only in the evening sky. It passes into Leo in the second week of June (passing 0°.4 N of *Regulus* on August 10). In mid-August it becomes too close to the Sun for observation until in the second week of September it reappears in the morning sky in Leo in which constellation it remains for the rest of the year. Its westward elongation gradually increases and by mid-December it can be seen for more than half the night. Jupiter is in conjunction with Venus on July 1, July 31 and October 26, with Mercury on August 7 and with Mars on October 17.

SATURN rises well before sunrise at the beginning of the year in Libra and moves into Scorpius in mid-January. It returns to Libra in the second week of May, is at opposition on May 23 when it can be seen throughout the night, and in mid-October returns to Scorpius. From mid-August until mid-November it can only be seen in the evening sky and then becomes too close to the Sun for observation. It reappears in mid-December in Ophiuchus and is visible only in the morning sky for the remainder of the year.

URANUS is visible at the beginning of the year in the evening sky in Pisces and remains in this constellation throughout the year. In mid-March it becomes too close to the Sun for observation and reappears in late April in the morning sky. Uranus is at opposition on October 12. Its eastward elongation gradually decreases and Uranus can be seen for more than half the night for the remainder of the year.

NEPTUNE is visible at the beginning of the year in the evening sky in Aquarius and remains in this constellation throughout the year. In the first week of February it becomes too close to the Sun for observation and reappears in mid-March in the morning sky. Neptune is at opposition on September 1 and from early December can only be seen in the evening sky.

DO NOT CONFUSE (1) Venus with Mercury in the first three weeks of January, with Mars in mid-February to early March and again in late October to mid-November and with Jupiter in late June to mid-July and again in late October; on all occasions Venus is the brighter object. (2) Jupiter with Mercury in early August and with Mars in October after the first week; on both occasions Jupiter is the brighter object.

VISIBILITY OF PLANETS IN MORNING AND EVENING TWILIGHT

	Morning		Evening	
Venus			January 1	– August 11
	August 20	– December 31		
Mars			January 1	– April 18
	August 6	– December 31		
Jupiter	January 1	– February 6	February 6	– August 13
	September 10	– December 31		
Saturn	January 1	– May 23	May 23	– November 13
	December 17	– December 31		

Agrarian Trust Principles: Preamble

At this moment in our history we are faced with an unreasonably high cost of land, and an unreasonably low price of food and agricultural labor. We aren't tracking the costs of environmental harm or health costs associated with our farming systems, which are increasingly under threat from a changing climate. We understand that we must shift normative standards that currently define American land use, which reward development over conservation, and extraction over stewardship.

In the next 20 years, 400 million acres of American farmland will change hands. Will it concentrate into the hands of agribusiness corporations and investors, or come under the stewardship and care of next-generation farmers producing food in a sustainable, regional economy ? Now is the time to grapple with the shift ahead, and to articulate a clear set of values about how we undertake the transition of land, wealth and power.

Our goal with the Agrarian Trust is to scale up the transfer of farmland to the next generation of sustainable farmers and ranchers. We proposed to act through internal efforts and external partnership and empowerment. To build the issue, strengthen the network of stakeholders (entering and retiring farmers, farm service providers) and to build model templates that meet the needs of the land and its caretakers.

We understand it to be in our collective self interest to support the kinds of economic and social relations that sustain the ecological integrity of farming. More than 60% of American farmland is leased, and much of it on a yearly basis, or on a handshake agreement; this does not empower farmers to be the best stewards of the land. Longer-term tenure is the norm in France and England, and from a historical perspective a "commons" approach to land-rights is far more reliable than the "fee-simple."

AGRARIANTRUST.ORG/PRINCIPLES

January

WE CAN TURN THIS AROUND

Soil

Chapter 1

An Essay On Soil

Connor Stedman

In Mediterranean oak woodlands of coastal California, decaying mounds and clumps of grass can be found scattered throughout the understory. They're easy to miss, even if you know what to look for. These old clumps are skeletons—the remains of bunchgrasses that once grew among the oaks. The bunchgrasses lived for decades, sometimes centuries, and sent roots deep into the soil to find and store water through the five to six month dry season. Before colonization by European-Americans, bunchgrasses lived in a complex ecological partnership with fire, elk, oaks, fungi, and native peoples, among countless other players.

During the Gold Rush, though, hillsides across California were stripped of trees, grasses, and soil alike in the search for precious metals. Then came cattle replacing elk and fire, and tillage farming for flax and wheat replacing bunchgrasses. Bacterial communities adapted to disturbed soils took over from the long-lived fungal communities of the oak woodlands. And while some oak forests have grown back, the bunchgrasses, elk, fungi, fire, and hands of the native peoples are mostly absent.

In the wetter Northeast, it can look like wild land has staged a comeback. After being mostly deforested in the early 19th century for sheep farms, forests have returned across the region after those farms were abandoned in the late 19th and early 20th centuries. The Northeast is once again dominated by trees, as it was prior to the colonial waves of clearcutting.

As those forest lands regrew, though, important members of the community didn't return. Passenger pigeons, that had once acted as a conveyer belt of nutrients up and down the Eastern Seaboard, were being hunted to extinction. Migrations of ocean fish up rivers, which had brought micronutrients and nitrogen from the sea inland for thousands of years, were reduced to a trickle by dams.

Spring ephemeral wildflowers, holding nutrients in the upper layers of soil against the leaching of spring melt and rain, were a fraction of their former numbers as their slow-to-disperse seeds struggled to return. Large carnivores too were suffering the same fates, which affected forests in less-obvious ways. Wolves' and mountain lions' predation on large herbivore herds caused these animals to alter their movements in the landscape unpredictably, resulting in a patchy quality to those herbivores' impacts on the forest. That diversity helped ensure that many different types of forest trees could regenerate and grow across the landscape.

There is a pattern here. Historically in each of these places, a web of complex relationships kept natural capital—carbon, nitrogen, mineral nutrients, water, biodiversity—in the landscape for as long as possible and in as many forms as possible. The simplifying and streamlining of the landscape by dams, roads, plows, impermeable surfaces, and a simplified food chain has quickened the loss of this natural capital.

California White Oak

We could tell a version of these stories for almost any bioregion—the forces of colonization and modern capitalism have ensured that the tale has been replayed over and over again. This narrative is well-known to many organic farmers who work carefully to keep water, topsoil, and biodiversity on the farm, but it's important for everyone involved with land to consider. We humans are the ultimate keystone species, after all, the fate of the entire living planet lies in our hands.

If we play our stewardship cards in the right way, in bioregion by bioregion all around the world we can regrow these ecological cycles and relationships. Farms,

forests, and grasslands can store and regenerate natural capital again, rebuilding the ecological fabric that is the ultimate source of our prosperity and survival.

But to know how to undertake that stewardship, it's not enough to know the land as it is now. We need to dig below the recent surface and go deeper—find the older ecological and cultural stories of a place. It's the wildlands that hold these stories, and it's these lands that will return them to us if we know where to look and how to listen. An agrarian economy needs to tend, restore, and engage in a deep relationship with the wild as well as with the planted field.

Sandbergs Bluegrass

Common Camas

Blue Oak

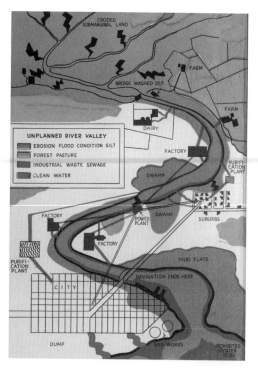

UNPLANNED RIVER VALLEY
ERODED SUBMARGINAL LAND
■ EROSION FLOOD CONDITION SILT
□ FOREST PASTURE
■ INDUSTRIAL WASTE SEWAGE
■ CLEAN WATER

ERODED SUBMARGINAL LAND
BRIDGE WASHED OUT
FARM
FARM
DAIRY
FACTORY
PURIFI-CATION PLANT
SWAMP
FACTORY
POWER PLANT
SWAMP
SUBURBS
FACTORY
MUD FLATS
PURIFI-CATION PLANT
NAVIGATION ENDS HERE
CITY
DUMP
GAS WORKS
PROHIBITED OYSTER BEDS

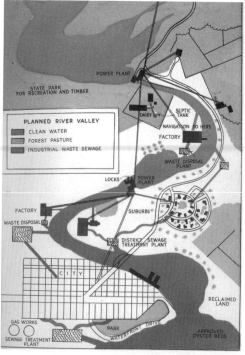

PLANNED RIVER VALLEY
□ CLEAN WATER
□ FOREST PASTURE
■ INDUSTRIAL WASTE SEWAGE

POWER PLANT
STATE PARK FOR RECREATION AND TIMBER
SEPTIC TANK
NAVIGATION TO HERE
DAIRY
FACTORY
WASTE DISPOSAL PLANT
LOCKS
POWER PLANT
FACTORY
WASTE DISPOSAL
SUBURBS
DISTRICT SEWAGE TREATMENT PLANT
CITY
RECLAIMED LAND
GAS WORKS
PARK WATERFRONT DRIVE
SEWAGE TREATMENT PLANT
APPROVED OYSTER BEDS

Fukuoka Knew

Steve Sprinkel

Iconic Japanese farmer Masunobu Fukuoka (1913-2008), patron saint of the No Weed Society (no affiliation with the US Drug Enforcement Agency) learned from Nature that conventional agriculture, sown in neat rows one after another, was an invitation for other plants of the undesirable variety to find a home between the cabbages. In One Straw Revolution, Fukuoka laid out a cropping system based on plants that create weed barrier mulches for successive plantings, and orders of succession designed to intimidate the unwanted. Missouri agronomist William Albrecht taught similarly.

We took what we could glean from Fukuoka's brazen faith as we observed truths revealed amid the waist-high jungles we manage. Why, we ask, do the worst weeds always grow in the carrot beds? Is it true what experts claim about weeds burgeoning when the moon is in opposition to Saturn? If I gather tons of nut sedge and roast it when Scorpio reigns, then sprinkle the ashes over my farm, will the sedge take the hint and be fearful of potential flame? If I plant the hell out of an area, will the intentional crop send out warning signals, not received by the human host, to every purslane and amaranthus in the vicinity, that the mighty kale rules beds 4, 5 and six in Block 4 of Section C?

I always doubted the efficacy of the theory. I enjoyed the mystical aspect and wanted to agree that Mankind, with our dedication to linear geometries and our desire to go straight, bought into the neat rows as merely another way to conquer an otherwise generous Nature.

The tractors tread in lines once carved by oxen, whipped and managed dead ahead, though the beast felt that it might be best to wander sideways with the contour because it was much less taxing. Later, observers learned that the rain partners with gravity, and loves to run straight downhill, carrying topsoil with it, all the way to the endless sea. Our best available block on the farm this fall was a fallow acre previously covered in onions, fairly well managed, not too overwhelmed with weeds, between the corn and summer squash.

On the fringes I had to mow down some ancient Swiss chard and cilantro, gone to seed I never had time to gather. I had scant faith in the chard seed and already own enough cilantro seed from previous crops to plant 20 acres. We watered the land up and bedded, threw down some fertilizer, tilled it flat in nice even rows and planted.

Seed-people are familiar with the vitality of fresh seed and judging from the carpet of unintentional crops that came up, their trust is well founded. You want chard? We got chard. Weeding commenced, when appropriate, on interior areas soon choked with malva, purslane and errant wild tomatillo. Each tomatillo will hold 84 seeds. There are 47 tomatillos on each plant and there had been 1,882 tomatillos bushes growing amid the chard nearest the corn. A seemingly benign little plant, nary a foot high, that, math be done, put out seven million four hundred and thirty thousand, one hundred and thirty six seeds. Over half the seeds sprouted in the first go-round, the others will arise no doubt once we turn the lovely soil again while weeding now, or for the next seven years.

For it is written: one year's sloth with the weeds will yield seven years of bent knees and broken hoe handles. The result is an unintended example of Fukuoka's Law. We saved the lettuce, with much labor, and in mid-weed determined to save the blushing chard plants and leave them for when the lettuce was harvested out. There was such a fierce blanket of cilantro amid the cabbage that almost nothing else came up, and Lord knows, there was plenty of weed seed available. But the cilantro ruled. It was like a 49er fan thinking he could go over to Oakland wearing some gold and red and watch his football team play the Raiders. That cilantro is the equivalent of the Black and Silver hordes monitoring the entry gates of the farm like toothy hounds, just daring a malva or some puny henbit to poke its head above the soil. The cilantro will be with you for some time, to enjoy its cleansing effects and occasional benefits in creating salsa with the tomatoes and chiles you still find. These greens are precursors of the fall that now is announced by rain today.

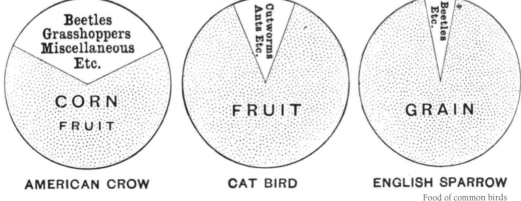

Food of common birds

SCARCITY VS. ABUNDANCE

THE FARM BLOC PIE

$4,100,000,000 of all FARM INCOME TO 25% OF U.S. FARMERS

ONLY 15% TO ALL THE LITTLE FARMERS

"FARM" BLOC

CONSUMER

SMALL FARMER

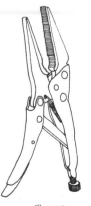

Big Farmers, advocates of scarcity and greater income for themselves through Price, are out to break through parity ceilings and increase farm prices 10 per cent. Producing at capacity themselves, they would prefer seeing small farmers and consumers plowed under than to see the adoption of a program for abundant production, which necessarily would fortify small farmers. They know that 25 per cent of farmers would get 85% of income from a PRICE increase and that the 75% of smaller farmers would benefit little.

Farmers Union advocates a program of providing facilities to small farmers so they can get into all-out production. Millions of them can double their production and their income at present prices, and get a larger share of an expanded farm income pie, while supplying consumers, the nation and the Allies with the foodstuffs they need.

Illustrations by Alana Rose

The Bio-Dynamic Rebirth Of Sun Dog Farms

Darby McCrea Smith

My husband, Elliot Smith, and I moved onto an abandoned farm in Blairsville, Georgia on January 13, 2013. We quickly set about establishing ourselves in one room of the small, aging farmhouse with a space heater, a camp stove, and a tub for washing dishes with creek water. It would be several weeks without running water, an indoor toilet, or an acceptable heat source. On top of these challenges, we were faced with the daunting task of starting up a new farming operation and removing all of the overgrowth and garbage that time had left in our path. This process required three roll-off dumpsters, hours of bamboo control, the purchase of a good used tractor, serious renovations to the house, and levels of patience I may never see again. There was only one reason to dedicate such time and effort to this farm after 10 years of abandonment, and that was the soil.

The property was formerly farmed by author and agricultural consultant Hugh Lovel, who wrote A Biodynamic Farm about his

work here. Lovel farmed the land for 30 consecutive years, turning this eroded mountain valley into a cohesive farm organism. He marketed his crops through a local CSA and hauled them to the up-and-coming farmers markets in Atlanta. Through his years of experience and research on the farm Hugh Lovel became an expert in the spiritual sciences of Biodynamic Farming. His partnerships with the Barefoot Farmer and other regional producers led to some of the first Biodynamically grown foods in the Southeastern United States. Lovel's attentiveness towards the holistic management of the soil, atmosphere, cosmic forces, and seasons created some of the most beautiful soil we had ever seen and it quickly became our inspiration.

After weeks of reclamation the bones of the farm organism were revealed to us. Our vision for what we could create began to come together. Terraces were uncovered, pastures were cleared of trees, a greenhouse constructed, and permanent beds

were resurrected from their fescue slumber. We added chickens, a cow, sheep, and some bantams for the yard. Just as the Spring crops set roots and the valley awoke with the promising hum of life, rain drops began to fall. They continued

throughout the Spring and Summer growing seasons, but in spite of our worries about flooding and erosion, this naturally-maintained landscape survived the torrential rains. Life carried on with grace.

Biodynamic growing has the beautiful underlying philosophy that everything is a reflection of everything else. A good example of this is found in the concept that the entire farm has similar needs to that of your own body. There must be effective circulation, elimination of toxins, sufficient amounts of water and nutrients, proper respiration and flow of oxygen, a sensitive nervous system, and the functions of digestion.

These attributes are derived from the influences of the celestial bodies, the presence of minerals, the formative energies of plants, and the transformative energies of animals. The fermented herbal remedies also termed the Biodynamic Preparations are used to harmonize the energies of these attributes. These Biodynamic Preparations aid improperly-functioning pieces of the farm organism using animal and plant matter that share energetic imprints. As an example take the use of Biodynamic Preparation 502 where Yarrow Flowers are fermented in the bladder of a deer.

This preparation is permeated by the energy of the planet Venus and utilizes Sulphur and Potassium to enhance the farm's ability to excrete waste and toxins, much like the same chemicals do in your own body. Both the Yarrow Flowers and the Deer bladder have energetic qualities that will enhance this process through their similar

relationship to Venus, Potassium, and Sulphur.

This replication of systems goes beyond the visible into the atmosphere and soil—each essential ecosystem working best when biodiversity and nutrient availability are high. Through total soil testing, slow release fertilizers such as granite dust and azomite, and the use of a Spader implement on our tractor, we have been able to maintain the quality of our soil while increasing the availability of nutrients. We have mulched as often and wherever we can, and focused on growing plants that remain in the soil long enough to build. We've kept and maintained Hugh's grass strip system between our permanent beds to keep erosion in check while preserving important soil structure.

We have utilized the healing forces of the Biodynamic Preparations, including the use of Barrel Compost, a fermentation that serves as an inoculant for good energy and biology in soil being prepared for cultivation.

All of these methods were inexpensive and far more effective than even the best OMRI approved fertilizers and mixes. Identifying the systems on your farm and giving them what they need to

function gives living things the best chance at a healthy life. This journey on this farm has brought us closer to everything. As human creatures on planet Earth we have the opportunity to acknowledge the beautiful relationships we share with everything from the Sun and the Moon to the microorganisms that bring life from death.

EDITORS NOTE An excellent method of weed control.

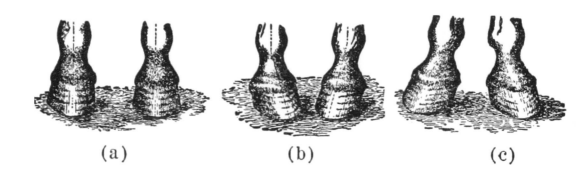

(a)   (b)   (c)

FARMING AS A FINE ART.

Farming is one of the fine arts. "The material out of which art is made is everywhere; but the artist appears only at intervals." The idea is prevalent among farmers that life must be a grind, a dull monotonous routine, devoid of beauty or enjoyment to all but the fortunate few. As soon as we realize that what we call life is opposite of this idea, the sooner will our farmers become artists. The farmer's work and life was not intended to be devoid of beauty. One's surroundings may be of the humblest sort, but we can look beyond these to the beauty of nature and the children's faces. Is the morning gray and chill in your home? Do the children feel this and act accordingly? Then search for some beautiful object near you with which to divert their minds.

Why not have in front of your dining room windows a beautiful vine or a clump of bushes. In winter why not fill the windows with blooming plants? By looking out you may see the brown leafless sunflower stalks, that seem to have nothing to recommend them; but watch a minute and there will come a flock of chattering sparrows eager for their breakfast, and it is a cheerful sight to see the stalks bend and sway, while our feathered friends satisfy their hunger by eating the brown seeds just ready to fall to the earth. Call the children to see the birds and observe how quickly the clouds will pass from their faces. The farmer's vocation is rapidly changing its tone, more and more does the farmer feel the sacredness of his calling and more abundantly does he love his life and work.

There is a difference between farming in its sordid sense and farming in its higher sense. First of all we need farming as a fine art as a direct contribution to life; the haphazard business man, the quack doctor and the go-lucky farmer are not good citizens. Thoughtful, earnest men have taken up the task of entering into the struggles of the farmer and the neglect of the farm to see if the incompleteness can not be, measurably at least, done away with. These problems of interest to farmers that are being so ably discussed by the students of these questions with the purpose of accomplishing results, are meeting with a cheering response from farmers; and those who have toiled for many years tell us that immense results have been accomplished in uplifting the general tone of farming, although the pessimistic farmers can see nothing but drudgery.

The Farmers Advocate is doing a great work toward making one of the fine arts, and the readers can greatly assist by standing for the highest and best in farming, and all can testify to your faith by your good work.

In connection with prevailing ideas concerning the curse of drudgery, two pictures come vividly to mind—one is "Holbein's Plowman," with its barren fields, its poor cabin, worn out horses, and the son ragged, cheerless and sad faced. This legend appears under the picture:

"By the sweat of thy face,
Thou shall maintain thy wretched life."

The very common notion of farm work, its misery and horror is here presented. The other picture is that of a bright October morning, when for the first time, a youth went to strike a furrow and to do a man's work. With the first rod the plow struck a rock and the boy tumbled headlong. By nightfall every muscle and bone ached, and the work did not merit much praise, but the youth's face was bright with hope and he was happy with the wholesome satisfaction that comes from achieving. There is an essential difference between the plowman and the happy boy; it was not in the kind of work, nor in the reward; the Plowman had his board and clothes; so had the boy; the surroundings of both were the same country life, country skies and country associations; but there is a marked difference. The two workers had been differently educated; the boy had been born and trained to Virgil's philosophy "Happy the man of the fields if he only knew it." While the plowman could gain no inspiration, no hope, no love for the work; the boy loved his work and the results of it were pleasures in themselves; it was a beautiful, sensible and satisfactory idea of life.

There has been and is still something wrong in educating our young people to feel and think the lot of the farmer a curse, and thus the young man or woman who remains on the farm must be cuddled and pitied.

The man who molds a beautiful statue of clay is an artist; so is the man who tills the soil in the right spirit. Taking care of cattle is as noble as climbing trees of knowledge, sowing seed if properly understood is on par with preaching a sermon; both have to be done in the right manner and in the right spirit to be among the fine arts.

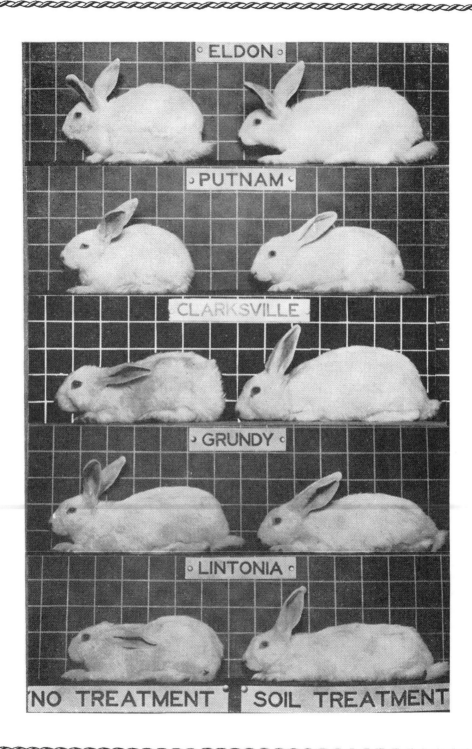

Willa Mamet *Field Fog*

Working With Natural Energy

Leaf Myczack

Webster's dictionary defines energy as "vigor in performance of an action… [and] vitality in expression." Nowhere is this definition more illuminated than in a healthy forest, or on well-managed farmland. The biological productivity of such land correlates directly to the consciousness and awareness of the land steward. As a farmer I work from experience when I interact with the inherent energetic forces that pervade the natural world. In a phrase, this means farming by being directed by the land energy itself.

Conventional wisdom holds that sudden changes in the physical structures that make up the biological web of life are inconsequential. And so we blithely clearcut whole forests, dam up rivers, plow under prairies, and smother rich bottomland under endless shopping malls and sub-divisions. For anyone with an ethics where the Earth environment does indeed matter, these actions represent a loss of natural energy, vigor and vitality in expression. Modern agriculture is foolishly ignoring that real is better than imitation—that real soil fertility is more likely to succeed than synthetic, and soil-toxic fertilizers; that natural plant health and disease resistance is more effective than toxic pesticides. Energetic imbalance in nature manifests as disease, and most of our current conventionally-produced food is diseased[1] and of low energy and nutritional value. It is a problem if domestic animal health is dependent on medications rather than local forage plants. It is a problem if the natural energy has been forcefully negated by human intervention in the pursuit of greed.

What is important to understand though, is that the sources of natural energy can be restored. However just as natural vigor was not suppressed overnight, there is no silver bullet that will restore authentic natural balance. It will take time, it will take commitment, and it will take a labor of love to accomplish.

Healthy land feeds healthy animals and people, and most importantly, feeds itself. Healthy land is vigorous land, and we feel the more vigorous and alive for sharing a presence together. Healthy land is the result of many species and processes working in symbiotic relationships with each other—processes such as rust, mold, fungi, wilting, and decay. Often these natural relationships are seen as a threat to mankind's edifices, structures, and vegetative manipulations and are targeted as nuisances

---

1   I am referring here to the fact that many of our commercially-produced crops are so dependent on synthetic disease-blocking agents that they cannot survive naturally without intervention by these toxic substances.

to destroy rather than essential forces in the regeneration of topsoil. This is short-sighted.

I can't actually measure the energy output of birds creating waves in our bird bath, the energy output of schools of fish swimming about in the farm ponds, the energy output of moles burrowing through the topsoil, the thousands of daily bee flights back and forth over garden and pasture. However I would be shortsighted if I ignored their impact on the intricate workings of our farm community.

Here is what clinches it for me: when I step back from my collaborative relationship with my farm community, there is no disruption in the natural flow of life. The energy, the vigor and vitality, is self-perpetuating. I am merely a willing worker in the big seasonal scheme, much like a bee gathering nectar, or a bird eating bugs. This way of being and acting as a respectful community member is the best way to benefit from the abundant natural energy at our fingertips.[2]

_____

2   This essay edited from its original version for the Greenhorns' Almanac 2015.

The Honorable Harvest

Know the ways of the ones who take care of you, so that you may take care of them.

Introduce yourself. Be accountable as the one who comes asking for life.

Ask permission before taking.

Abide by the answer.

Never take the first. Never take the last.

Take only what you need.

Take only that which is given.

Never take more than half. Leave some for others.

Harvest in a way that minimizes harm.

Use the harvest respectfully. Never waste what you have taken.

Share.

Give thanks for what you have been given.

Give a gift, in reciprocity for what you have taken.

Sustain the ones who sustain you and the earth will last forever.

—Robin Wall Kimmerer, from *Braiding Sweetgrass*

Illustration by Josef Beery

EX LIBRIS

NORTH CAROLINA

CONSTANT

FAITHFUL

THE SANDHILL
FARM-LIFE SCHOOL
Gift of

R. O. Palmstrom, '21

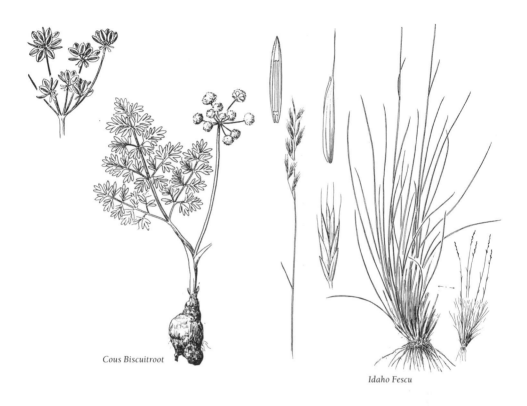

Cous Biscuitroot

Idaho Fescu

28-CENT MACHINE GROWN WHEAT WILL DESTROY AMERICAN FARMER

Wichita, Kans.—Corporation farming in this State can produce wheat at less than 28 cents a bushel, and this price will pauperize the American farmer, declared Harold McGugin, Kansas congressman-elect in a radio address.

"Unless corporation farming is stopped all national agitation for farm relief will have been in vain," said Mr. McGugin. "Corporation farming, not bearing the expense of maintaining families, produces its wheat very cheaply. These corporations are breaking up thousands of acres of new land. They increase acreage thereby increase the wheat surplus. They will continue to do so as long as the price of wheat will permit a profit over 28 cents a bushel.

"Corporation farming will solve the farm problem, but it will impoverish the individual American farmer and drive millions of them into the labor centers of America only to increase the present economic chaos.

Lucas Foglia *Tommy Trying To Shoot Coyotes*

THE FARMERS ADVOCATE

Volume 28. No. 49.   TOPEKA, KANSAS.   July 12, 1906.

SOME ADVANTAGES IN MULES.

Those who are accustomed to handling mules and know their remarkable faculty for adapting themselves to almost any condition, realize their value and do not hesitate to pay well for good animals, says Rural Home.

The advantages they possess over horses in many ways compensate for the somewhat objectionable length of ear and exaggerated reputation for kicking. They are ready to work when two years old, and on account of their toughness and steadiness of nerve may be kept at it twice as long as horses are fit for service. Mules can be raised more cheaply than any other stock, and, unlike horses, will never founder themselves if by chance they should have access to more feed than is good for them. They can be put into the market much sooner than horses, and if properly handled, when two years old, will do as much work, and stand it much better, than 4-year-old colts. Mules are less liable to contract diseases than horses and their value is not decreased by blemishes as is that of horses. Mules seldom run away, but when they do they seem to run more from sport than fright and generally wind up safely. They instinctively avoid holes, obstacles or dangerous places. They can stand heat, abuse or hardship better than horses and can always be relied upon. They are sensitive as well as sensible animals, responding quickly to kind treatment by docility and gentleness.

The great profit in mule raising lies largely in their growth. Mules grow so quickly that they are marketable when three years old, while horse colts can not advantageously be sold under five years.

Buyers will always handle good mules, as the demand for them is steady. Where large numbers of mules are raised there are always plenty of mule buyers in that section of the country. Buyers do not go where there are no mules to sell. In times of war mules are always in great demand and bring good prices owing to their powers of endurance.

Comparing cattle with mules, one steer will eat as much as two mules. A good steer at the age of three years is worth from $75 to $100, while a pair of good mules the same age will sell for $200 to $250 without consuming any more food. As compared with swine, if the same amount of feed that is fed to a bunch of hogs is fed to mules it will not take long to find which make the most money, aside from the fact that there is no risk from cholera.

In orchards or nurseries mules are invaluable. No animal can equal them for running out straight lines, and in cultivating young trees they seldom break or injure them. When well trained they are careful and obedient.

Mules occupy a place among our domestic animals that is indicative of honesty, durability and valuable service, and their failure to breed is looked upon by many as an indication of the fact that in them the climax of perfection has been reached.

Who Speaks For the Farmers?

By Helen Fuller, in The New Republic

The Byrd Economy Committee of Congress has been busy recently trying to make certain that the tenant farmers and sharecroppers of Southern poll-tax states are kept as voteless as they are now. The committee has been looking into charges brought before it by the American Farm Bureau Federation and Probate Judge Bob Green of Hale County, Alabama, that the Farm Security Administration has been paying poll taxes for some of its clients in Alabama.

When Senator Byrd called his fellow Virginian, C. B. Baldwin, Federal Security Administrator, before the committee, Mr. Baldwin testified that the FSA was approving loans to clients who needed money to pay present or back poll taxes. In Alabama, poll taxes are cumulative up to $36, which is a very large sum to farmers of whom more than half are still earning less than $500 per year. Baldwin also said that the FSA was charged with rehabilitating those farm families to which it was making loans and that the FSA considered full voting status an essential part of such rehabilitation. The Byrd Committee's all-star cast of Senators from poll-tax states were horrified by such brazen belief in constitutional guarantees, and Carter Glass was quick to assert that if the Department of Agriculture solicitor agreed that such procedure was legal, Mr. Baldwin had better get himself a new solicitor.

Senators McKeller of Tennessee and George of Georgia made equally determined noises to indicate that tenants and croppers are not going to be forced to own their own farms, paint their houses or feed their children balanced diets so long as they are in the Senate to ward off such a fate. Some cynics attending these hearings pretended to see a connection between the attitude displayed by the four senior statesmen from the South on the committees and their personal electoral problems. They pointed out that at their last reelections, Senatoss McKellar and Byrd were returned to the Senate by the votes of approximately 17 percent of the population over twenty-one in their respective states; that 15 percent of adult Virginians were enough to send Carter Glass back to his old seat in Washington; and that in the off-presidential year of

1938, when he last ran, Walter George came in first with 3 percent of Georgians over twenty-one casting votes for him.

More seasoned Washington hearing-goers could think of further explanations for the Byrd Committee's all-out attack on the farm-security program. They recalled other occasions on which Ed O'Neal, Farm Bureau Federation president, has played a star role.

They remembered the time last year when Ed O'Neal and his organization declared war against the Department of Agriculture and especially against their chosen enemy the FSA, sometimes referred to as "the poor man's Extension Service." Their suggestion then was that the FSA be abolished and its functions divided between the Farm Credit Administration and the Extension Service of the Department. O'Neal and his boys lost that round, but there is very little reason to believe that they accepted the defeat as final. The Farm Bureau is now making the same recommendations before the Byrd Committee. Perhaps it is time to ask why the Farm Bureau is so persistent in its attack on the FSA. What is the Farm Bureau Federation anyway?

The first chapter in the history of the American Farm Bureau Federation was written in Binghampton, New York, in 1909—prophetically enough, under the sponsorship of the Binghamton Chamber of Commerce. From that time to this the Farm Bureau has consistently stood for the interests of the well-to-do farmer and his friends in industry. Baldwin was safe in saying, as he said before the Byrd committee the othe rday, that "Mr. O'Neal cannot by any stretch of the imagination be considered a representative of the low-income farmers." Essentially, Farm Bureau philosophy has followed the school of thought which holds that "poor farmers are poor because they are lazy."

In its first stages, the Farm Bureau operated on a county basis to spread knowledge and practice of scientific-farming methods, as they were developed by agricultural departments of the land-grant colleges which had been established by westward-looking settlers. At a very early period, Farm

Bureau activities were closely interwoven with those of the Extension Services of the land-grant colleges, so that it was quite natural, when the Smith-Lever Act of 1914 came along, for the Farm Bureau to take rapid advantage of its benefits. The Smith-Lever Act provided grants for extension work to states which were willing to match federal funds, either with state, county or local funds or with funds supplied by "interested groups of citizens." Because the Farm Bureau was financially able and quite willing to answer to this last description, the American Farm Bureau Federation, with scarcely any effort at all, was able to obtain virtual control of our national agricultural machinery. And after more than twenty-five years, it has yet to be dislodged from that position. By supplying the matching funds called for in this early act, the Farm Bureau controls farm organization in eleven key states. Now, by contributing approximately a million dollars out of a total annual expenditure of close to thirty-three million dollars for extension work throughout the country, a private organization thus is able to control a public service.

Well-to-do farmers quickly realized the advantages which could come from close contact with the state agricultural colleges, and by the time of World War I, they had moved to consolidate their control of these institutions. The World War put the Farm Bureau on the map, calling as it did for lightning-like expansion of the infant Extension Service. Following the war, the bureau adopted a legislative program and began to look far beyond the field of scientific agriculture. By 1922, there were probably 1,250,000 Farm Bureau members and the organization was ready to get into big-time lobbying company in Washington. It succeeded. A good indication of its progress can be seen from the important role the Farm Bureau Washington lobby, headed by Chester Gray, played in determining the disposition of Muscle Shoals after the war.

By 1928, Chester Gray had established the Farm Bureau as a lobbying factor that congressmen had to reckon with. By that time, the big-business sympathies of the bureau had been clarified in several election campaigns, and it came as no surprise to men like Senator Norris to find the Farm Bureau fighting against the real interests of American farmers on the question of the disposition of Muscle Shoals. This was a long and bitter struggle. Farmers had more to gain by proper disposition of the Shoals than almost any other group, yet by intrigue and deception, Farm Bureau representatives were able to throw the weight of the organization over to the side of the power lobby, which had every intention of fleecing the farmer as shortly as possible.

As the Coolidge and Hoover regimes slid off into history, the Farm Bureau began to suffer real losses in membership and standing. By 1934, it was down to some four hundred thousand members and lacked the kind of program needed to attract the dissident farmers of the period. The Farmers Union began to gather strength, and very early in the New Deal farm problems seemed to divide into a three-way proposition: wheat, corn and cotton. The Farmers Union was soon established as the spokesman for wheat. Farm Bureau strength had long centered in and unquestionably dominated the corn country. The Cotton South remained virtually unorganized. The Farm Bureau in 1933 had practically no strength in the South. The next move was obvious to politically astute Farm Bureau bosses like Earl Smith of Illinois, who had headed the organization as long as a corn boss could do the job. As soon as he understood the score, Smith looked around for a likely cotton man to front for him. Ed O'Neal, then a national vice-president, seemed made to order.

If you have ever seen Ed O'Neal you know that Earl Smith was a good picker. Even if you have not seen him, his record since 1931 proves that Smith was right. When the New Deal came in with its newfangled notions of how to make American agriculture a paying proposition, the Farm Bureau was in a more strategic position than its declining membership rate indicated. From years of infiltration and careful organization,

the Farm Bureau had a corner on most of the trained professionals in the agricultural field. When the AAA and the rural-rehabilitation programs came along, their choices of field personnel were fairly well limited to those trained by agricultural colleges, and consequently in all probability touched to some extent by Farm Bureau philosophy and organizational controls.

Regardless of how good programs were when they came out of the Washington office, they were translated on the operating level into a Farm Bureau version of the original idea. But in spite of these basic controls, the Farm Bureau needed to bolster its membership, so someone thought of adapting the check-off principle to farm organization. In many states it has worked like this. Big planters who have always been Farm Bureau members and understand that the bureau is working for their interests have agreed to sign up all their tenants and croppers for the bureau. The member rarely knows he has joined; his two-dollar dues are just deducted fom his crop or his AAA check and turned over to the bureau directly. Under this ingenious system, membership in the South has shot up, and the Georges, Byrds, Glasses and McKellars have an equivalent respect for the Farm Bureau interpretation of agricultural economics.

As things stand now, Ed O'Neal and the five-thousand-odd members he has now are lined up against the FSA and its six hundred thousand low-income farm families. O'Neal's well-to-do planters are producing almost to capacity at present. If we are to meet the requirements of our Victory Food Program, it will have to be through increasing production by the three million farm families who now fall in the under $600 annual-income group. It is the Farm Security program which is capable of doing this—not the Farm Bureau program. It is our Baldwins who can win the war for us; our O'Neals who can lose it..

The Climate Resilient Farmer

Eli Rogosa

How landrace bio-diversity can feed a planet facing unprecedented climate change or why is modern wheat making us sick?

Our beliefs about our past shape our understanding of who we are as human beings and farmers, of how our foodcrops evolved, and of our potential to co-evolve with plants today. Just as the landrace seeds are almost lost, the practices that evolved the landraces themselves are forgotten as well.

Let us rediscover the forgotten seeds of biodiversity that have been generously gifted to us by generations of peasant farmers. Let us relearn age-old traditions of polyculture that create flexible, complex and resilient farms and gardens that can better adapt to unprecedented weather extremes.

Modern wheat is the most widely-grown crop on the planet, and like a canary in the coal mine it sounds the alarm about the industrial culture that developed it. It is bred with dependence on petroleum-based agrochemicals to survive. The roots are short and stunted for easy nutrient uptake of synthetic agro-vitamins. They are dwarfed

to less than 2 feet tall so that the plant will not collapse under heavy nitrogen applications. This height also allows for ease of harvest by goliath combines because the plants are now too short to scythe by hand. Modern wheats make us sick, their indigestible gluten laden with toxic nitrates from chemical fertilizers. Bred for total uniformity and enslaved, they are unable to adapt to new climate conditions. The seeds are patented to prevent farmers from saving them.

Ancient wheat landraces evolved in organic fields in the hands of traditional farmers. They developed from wild wheat over millennia to be uniquely well-adapted to diverse micro-climates of rural villages—valleys or mountains, parched deserts or rainy coastal fields. Generation by generation, seed-savers, most often women, selected the plant survivors of adversity for robust health and rich flavor. Landrace crops have deep root systems that reach down to the lower soil levels for micronutrients

and greater soil moisture during drought. Extensive root systems hold the plant steady during torrential rains. Tall, majestic, and flexible, landrace wheats are powered by sunlight instead of fossil fuels. They are rich in healthy plant-based phyto-nutrients and trace minerals that are easy for humans to digest.

Landrace wheat inspired awe and reverence in the peasants who grew it and believed that the grain field is animated by a soul like that of humankind. The life processes of reproduction, growth, death, and decay are the same principles that manifest in human beings. These past wheat farmers believed that in the plant, as in the human, there is a vital element—that the soul of a plant is like the vital soul of the human. This belief in the plant-soul is at the heart of age-old cereals traditions. The spirit of the grain was known as the Great Mother.

The Bread Hearth was the principal feature of ancient temples throughout Old Europe. Shrines with female figurines grinding flour, kneading dough, and baking bread are found everywhere in these spaces. The bread oven was a sacred space of the Great Mother. This can be seen in the many anthropomorphized 'womb' ovens found in these temples. These ovens are often in the shape of a pregnant woman with an umbilical cord on the top and engraved with energetic spiral lines.

Dough prepared in the Temples was sacred bread used in life-affirming worship. Breads marked with the 'spirals of life' were probably the first offerings to the Earth Fertility Goddess, combined with folk planting and harvest traditions that celebrated the bounty of the earth.

Let us rededicate our fields and tables to the life-affirming vitality and generosity evoked by landrace wheat!

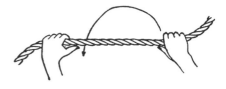

Jacob Cowgill

February

Servitude

Chapter 2

Nettle Story

Patrick Kiley

I am crawling all over with what feels like narcotic withdrawal but it's just nettles. I am tired but I can't sleep because I want to rub my skin off. It's midnight. There are heightened emotions living across my skin caused by plant toxins that I underestimated.

This is the first time I have felt nettle tearing through the mind-body membrane and waving its poisons like tasers. It is 12:10 AM and I am alone in this room with the fan running and an old window screen loaded with drying nettles. This is far, far from drunkenness.

Does magic experience phase states, as it leaps from one body to another? Maybe it is a phase state itself?

Nettles have serrated heart-shaped leaves that grow in pairs on a tall thin stalk. To harvest you take clippers or scissors and cut just above one of the plant's axils (leaf pairs), which will allow it to keep growing. The axils alternate in their orientation, climbing the stalk opposite each other. The plant grows quickly to be as high as a person.

The complex emotion I feel is both nausea at the thought of nettles and uncontrollable attraction to them. My body is swinging back and forth between two possibilities: either become nettles or be a person who does not disturb the nettle corpus. The whole extended family is singing to me a song of warning and welcome.

I have to go to sleep somehow. Tomorrow I have to be a person. Can I work this out in my dreams?

Nettles are good for your adrenal glands. They are very high in iron, calcium, potassium, vitamin C, and other nutrients. They help women's fertility and reproductive cycles. They grow in groups always, sometimes in an unquantifiable potent assemblage like an army at camp, dense and homogenous. At other times they are more like groups of strays just finding each other after a battle. I feel like the object of tactical violence.

I say it is the first time I've felt this way because in the past I took the tingling sting of nettle hairs as a sort of feather tickle. It is said to be good for one's circulation, and I believe it. It doesn't hurt unless you do it a lot. Then it hurts.

It hurts so much that I am short-tempered and mad with the pain. I am too incapacitated even to seek the soft kiss of jewelweed, no doubt growing within several hundred feet of me. But I cannot get it because I think going

Stinging Nettle

cutting off my hands if they'd grow back tomorrow.

This agony is exceptional in that, as bad as it feels, I am tempted to have more. This particular sort of pain, euphoric nettle pain, is calling for greater expression through me.

Some kind of subtle and messy reproduction is taking place here, a truly wild asexual folly. I'm cloaked in nettle, or it in me, and I'm helping it do its work—not the bacteriophagic work of consumption but something more like education. Conversion.

The harvesting earlier today was like this as well, I see now. I continued in my quiet exertions because I felt I needed more, and more was at hand. We were in dappled sun and shade, which is ideal both for the nettle and myself. The feeling of furtive persuasion and beckoning began at the outset and is refracted inward now in the most intense phase.

outside now—1:15 am—would only make this worse. To stumble around Hudson in my socks and underwear looking for jewelweed in Prison Alley: too extravagantly human.

It hurts so much that I am writing. I am lucid and stimulated and upset. I want to put my feet into damp woodland dirt and feel its coolness. I want my own small poison hairs and a chlorophyll heartbeat that I share with a thousand fellow beings. I just tried to punch the cement wall in the dark. I would take the quick pain of

I long for this: The jewelweed's watery stem breaks open releasing a cool wave allaying the wicked fire of acids and histamines injected by the needling trichomes.

2:30 AM. No real sensation goes unshared. The nettle and I sit across from each other in perfect disagreement, its leaves all fallen and pooled at the base of its naked stem. I regard this thin green thing and out of a deep and tutored love refuse to touch it with my bare hands.

Goodnight.

Senate Will Probe Evils of Farm Bureau-Extension Service Tieup

(Your Assistance Is Needed)

Thorough hearings on the improper relationships of the Extension Service with the Farm Bureau are planned by the Senate Agriculture Committee in connection with Senator Thomas' bill to divorce the governmental agency and private farm organizations.

Investigations will be conducted either in Washington, ir various states where abuses are common, or both. Kansas has been suggested as the location of one of the hearings.

Kansas Farmers Union is now compiling its file of evidences of violations. We need the assistance of every member of the Farmers Union and its friends to be sure that the subject is fully presented.

How You Can Help

Do you know of any cases of improper conduct? Of county agents declining service to non-Farm Bureau members, buying and selling goods or insurance for Farm Bureau, conducting membership campaigns, editing Farm Bureau columns or publications?

The regulation of the U. S. Department of Agriculture which is widely violated follows:

"As they (county extension agents) are public teachers, it is not a part of the official duties of extension agents to perform for individual farmers or for organizations the actual operations of production, marketing, or other various activities necessary to the proper conduct of business or social organizations.

"They may not properly act as organizers for farmers'\ associations; conduct membership campaigns; solicit membership, edit organization publications; manage co-operative business enterprises; engage in commercial activities; act as financial or business agents, or take part in any of the work of farmers' organizations, or of any individual farmer, which is outside of their duties as defined by law and by the approved projects governing their work . . ."

Please read the regulation carefully. Then, if you know of a violation of any of the provisions of the rule, please write a letter about it to President E. K. Dean, Kansas Farmers Union, Box 296, Salina, Kansas.

Has your county agent participated in Farm Bureau organization work?

Does he sell serums, insurance, seeds or other commercial products?

Do you know of neighbors or friends who have information about misconduct?

You can be of tremendous help to your Farmers Union by taking time IMMEDIATELY to report all the information on the subject you have, giving us the names of people who know and can tell about such episodes.

Act now. Write:

E. K. Dean, President,
Kansas Farmers Union,
P. O. Box 296, Salina, Kansas

"It is obvious that the bare land with its contents and the waters that flow through and about it constitute the nature-provided environment of human beings and are rightly the subject of their equal claims. Also that the value-for-use of these natural resources is conditioned on population. It follows population as its shadow. It appears with the people and disappears when they go. This value, therefore, should, by the best of titles, be retained by the community as its most excellent source of public revenue. The more the community draws upon this vast, community-conditioned fund the less will be the forced contributions from labour and capital. This means that the greater and better distributed will be the purchasing power of the people."
—Herbert J. Davenport

The American Farm Bureau Federation

WHAT IS IT?

1. The only Farmers' organization originated and financed by the U. S. Government.

2. The only non-sectarian, non-secret national organization of farmers.

3. An organization representing 31 state organization of farm bureaus.

4. The only organization with influence, prestige and authority to speak for farmers as a whole.

5. The only organization offering the farmers a referendum on all state and national questions.

WHAT IS ITS OBJECT?

1. Efficiency in production.

2. Economy in transportation and marketing.

3. Influencing educational, legislative and economic policies.

WHAT DOES IT COST?

Nothing compared to what the laborer, the packer, the merchant pays for his organization.

Nothing compared to what other states are paying for membership in the farm bureau. Iowa, Indiana and Michigan farmers pay $5. Illinois farmers pay $10 to $15.

The Pawnee County Farm Bureau fee is only $2 per year.

Every Pawnee County farmer should be a member of this organization.

The Pawnee County Farm Bureau

FARM TENANTRY IN EXISTENCE A VERY LONG TIME

Not Much Difference in the Present Form of Leases and That of Centuries Ago. Four Ways of Dealing With the Tenant Problem in the Past

The tenancy problem in America demands serious consideration, in the opinion of Dr. H. J. Waters, president of the Kansas State Agricultural college, who spoke on "the Landlord and His Tenant" before a Farm and Home week audience Tuesday evening. Moses, in ancient times, had the best system in dealing with the problem, while the United States has few plans of any sort for dealing with the situation.

"Americans get excited over the increase in tenancy every time a new census is taken and it is well that this is the case," said Doctor Waters, "because of the evils of our system of farming.

Back to Ancient Day.

"It is a mistake to assume we are the first people to have to meet this sort of problem. The fact is that tenancy dates back to the beginning of the established order of agriculture, before man used typewriters for writing farm leases or fountain pens for signing them.

"The earliest leases of which there is record were written in clay and sun dried, and the signature made by thumb nail impressions, but the terms of the leases were exactly the same as those of today—one-half the crop when the oxen and seed were furnished by the landlord and one-fourth to one-third when the tenant furnished the seed and the work animals.

As Moses Established It.

"Moses established the best system of land ownership that has been known, intended definitely to prevent land monopoly and to keep farms the size that would be best for everybody. One important provision was that the land belonged to the Lord and the one who tilled it held it for a rental of two tithes to support the priesthood, which had no land, and one tithe for the public welfare. The tenant could hold this land only 50 years at most or until the year of jubilees, when all contracts ceased and all debts were cancelled. If we redistribute the land now once every 50 years, we should hear little complaint of oppressive landlordism."

Long experience has shown that the system of land tenure which best saves the land from excessive wear and waste is for it to be tilled by the owner, pointed out Doctor Waters. Even greater protection was afforded to the soil under the old Jewish system whereby the land was considered to belong to God and it was sacrilege to despoil it.

Dealing With Tenant Problem.

Moses was a believer in small farms tilled by the owner. Gradually this system broke down, estates became larger than the owner could cultivate to advantage and a system of landlordism and tenacy again came in.

"There have been four ways of dealing with the tenant problem in the past," said President Waters. "Germany stepped over from the feudal system to the system of small farms without the intervention of landlord and tenant and as a result it may be truthfully said that Germany never had a real tenant problem. This is the only conspicuous example of preventing tenancy.

"France broke down the system of

land monopoly and tenancy by means of the French revolution and the long years of readjustment following, and as a result France today practically has no tenancy problem, but instead one of the most prosperous and successful land owning systems of agriculture in the world. Denmark accomplished the same result more recently without bloodshed.

"New Zealand is an example of a nation that thus far has prevented tenancy. The motto of her nation is, 'The land for the people,' and a system of graduated taxes makes it unprofitable to hold land not tilled. In other words, New Zealand has taken all speculative values out of land.

"Great Britain stepped from feudalism to tenancy and instead of trying to break it up, accepted and regulated it. As a result, Great Britain has the best regulated system known in the world."

The United States, Doctor Waters showed, has done nothing to prevent and almost nothing to regulate tenancy. In Great Britain nine out of 10 of the farms are tilled by tenants—in the United States about four out of 10, in France one out of ten. In New Zealand the number is negligible. The problem of readjusting a system of land holding in which three-fourths of the farms are tilled by the owners is a difficult one, and in a democracy such as ours, is perhaps a good ways ahead of us.

"In general the tenant has been a soil robber and waster because he has less interest in the land than if he owned it and because he is as a rule a crop farmer and not a live stock man," said Doctor Waters.

"In America the tenant has wasted the soil more rapidly than in any other country because he has been provided through American invention and genius with tools and machinery by which he can till more land than any other tenant and because under our system of short leases we encourage the most destructive system of farming known in this country or in any other country.

"English farms are tilled almost wholly by tenants and yet English soil has steadily increased in fertility during the last 100 years. Live stock farming has reached its highest development in England on farms tilled by tenants.

"A rented farm in this country means a run down farm. It is a farm with poor buildings, few fences, and no conveniences. The tenant could not, if he would, keep live stock. He

coudn't afford to rotate crops and intersperce alfalfa, clover, and cow peas with corn, wheat, and oats, even if he wanted to do so. He will not be there next year. What difference does it make to him how much the fields are washed or worn through carelessness and neglect?

"Generally, successful men who have been able to accumulate a competency move to town. They know the soil of their farms as no one else knows it. They usually take an active interest in the management of the farm by the tenant, encourage him to get good seed, help him plan his cropping system and his work so that the land is well prepared and the seeding done on time.

"But in the course of time this experienced farmer and landlord passes away and the farm is divided among the heirs, one of whom is likely to be the local banker, another the superintendent of schools in a city 100 miles away, and the third, the wife of the pastor of the church in another county or of a local merchant, lawyer, or physician.

"These new proprietors know little about farming except what they remember of their earlier farm experiences, and they are absorbed with their own problems and duties. Under our system of leasing the tenants who received such careful and valuable instruction from the older farmer have long since moved away and a new man, who is wholly unacquainted with the land and perhaps of limited farm and business experience, is on the land directed by these inexperienced and otherwise busy landowners. The result is bound to be a relatively low income.

"Tenancy is in many respects an evil, but if evil, it is not all evil, for there are good tenants as well as poor ones. Then, under our present system of land tenure, tenancy is necessary and inevitable.

"Live stock farming does not favor the development of the tenant system and where live stock is generally grown in this country few tenants are found. Where the tenants comes into live stock regions the herds are soon dispersed, the pastures are plowed and planted to grain, and the barns and fences fall into decay.

In England under a system of tenancy, live stock farming has reached a higher degree of development than in any other country. But the system of leasing in England is very different from that prevailing in this country.

"In every country of Europe land has been improved in fertility within the last half century. In this period we have wasted the American soil at a rate far beyond that of any other people or any people in any other age.

"This has been due partly to the fact that we have had labor-saving, efficient machinery with which to till our soil and in part to the fact that the American farm had to be cleared, paid for, and improved out of the soil. and for the most part within this period. The quickest and surest way to raise money with which to meet the interest and principal of a mortgage or with which to build a home, barn, fences, silos ,and windmills is to plow the life out of the land."

We Want The Whole Loaf!
Local, Organic And Fair

Elizabeth Henderson

Sebastiao Salgado

With the financial recovery looking more like the Great Recession, people are turning to the real goods and services of the earth economy. As stock prices rise and the top 1% bloats with wealth, for many in the 99% incomes are eroding and job security is a quaint concept from the past. A stream of books by Michael Pollan, Eric Schlosser, Bill McKibben, Barbara Kingsolver, and others have put local food at the top of the best-seller list. And deeper, systemic analyses (Wayne Roberts' *No Nonsense Guide to World Food*, Wenonah Hauter's *Foodopoly*) are helping us understand why the hard times are failing to evaporate and how small local actions can add up to transformative change. Stressed families are eating out less and planting gardens. Small seed companies are experiencing double digit growth. The number of CSAs all over the country has tripled over the past ten years. Downsized bankers and PhDs are signing up with the spreading network of new farmer programs. And more people are turning to local farms for the ingredients essential to their newly-recognized priorities of health and self-reliance.

The pundits tell us that local trumps organic—that for many people a certified organic label from a farm 3,000 miles away does not provide the reassurance they seek. Will they be satisfied when they realize that the farms within 100 miles of home use toxic chemicals, underpay their workers, and can barely manage to stay in business? People need more from farms than comforting proximity.

Farms that are certified organic wash their produce in drinking-quality water that is tested annually. Use of raw manure is carefully regulated and must be incorporated in the soil 120 days before harvesting a crop that touches the soil. In addition there is research that shows that a biologically active soil is better at eliminating pathogens than a fumigated soil with the life knocked out of it.

Our smaller scale and the required audit trails for organic certification ensure that we can trace any food people buy from us. We will need our customers' help preventing the government from strapping our local farms with food safety measures designed for industrial-scale problems.

Food safety is important, but there are other qualities which are necessary for an enduring food supply as well. If buying from a trusted farmer is to be more than a passing fad, we will have to turn away from the quick-profit mentality of the financialized global economy and rebuild local food systems. The first requirement for a lasting agriculture is fertile soil. Healthy soil is the foundation upon which everything else rests.

Fertile soils teem with biological life that thrives when tilth is excellent and the mineral content is well balanced. The people who plant in these soils must be skilled in maintaining fertility and protecting the soils from erosion and contamination. Public policy must support efforts to rid soil, water, and air of toxic pollutants.

In this era of industrial agriculture farmers have neglected soil health, and treated soils as a medium to manipulate with heavy equipment and to douse with chemical fertilizers. According to USDA food nutrient studies, the result has been a steady decline in the nutritional value of vegetables, fruits, and grains. By contrast, recent studies of organically grown foods show that in addition to lower levels of pesticide residues, organic foods have higher levels of vitamins, minerals, and antioxidants than conventionally grown foods..

You may pay a few pennies more per pound, but you will save in the long run through reduced payments for medical expenses. Nutrient-dense foods produced in season by local farms provide health care—a wiser strategy than buying medical care.

Becoming a locavore follows a deeper change
in consciousness, and provides countless benefits:

YOUR MONEY IS CIRCULATING
IN YOUR OWN COMMUNITY

Family-scale farms are independent businesses
that tend to support other local businesses,
and no corporation is siphoning-off your dollars
to line its coffers or pay its stockholders.

EMPLOYMENT IN LOCAL AND SUSTAINABLE
AGRICULTURE IS AS CLOSE TO THE DEFINITION
OF "GREEN" JOBS AS WE MAY FIND

ECONOMICALLY VIABLE FARMS PRESERVE
OPEN SPACE AND BEAUTIFUL LANDSCAPES

EATING LOCAL FOOD SAVES ENERGY

David Pimentel, Professor of Insect Ecology at Cornell University,
has calculated that modern industrial agriculture expends 10
calories for each food-calorie produced. Many of those excess
calories are burned up in transportation, packaging, and marketing.

BUYING FROM FARMERS YOU KNOW IS ALSO
A GOOD WAY TO ENSURE FOOD SAFETY

The source of your food is not lost in an endless regression of
distant middlemen and brokers. As farmers, growing for market is
an extension of growing for our own families.

—Elizabeth Henderson

Farming with organic methods can also help combat global warming. Research from the Rodale Institute has shown that if organic methods were applied to all of the world's cropland, it would be the equivalent of taking 1.5 billion cars off the roads. 39% of the world's carbon would be sequestered in the soil! By using cover crops, composting, rotating, and mulching, organic farms can sequester 1,000 lbs of carbon per acre per year. Building organic matter buffers soils, and in extreme conditions of drought and flooding organic farms are more resilient and out yield farms using synthetic fertilizers. Pimentel's research shows that on average, organic farms use 30% less energy than conventional farms by avoiding synthetic nitrogen fertilizers and focusing on recycling nutrients. The production of synthetic nitrogen fertilizers is especially energy-intensive. Evidence is surfacing that as much as 5% of nitrogen fertilizer turns into nitrous oxide, a greenhouse gas 300 times more potent than carbon dioxide.

For local eating to remain possible our local economy has to be economically viable, balancing food access for everyone with prices that allow farms and other food establishments to be decent places to work. Along with clean air and water, food is a basic human right. This has been recognized in the Universal Declaration of Human Rights and by every civil government except our own. Until we build a movement with the power to create full employment with decent wages for all citizens, we will need to transform the agricultural subsidies currently lavished on agribusiness into mechanisms that make food affordable for those with low incomes. As Eric Schlosser's *Fast Food Nation* has made clear, most of the people working in our cheap food system are

Ginny Maki

underpaid, overworked, and often undocumented. Not only do farmworkers earn some of the lowest wages we offer, they are also not covered by the National Labor Relations Act. Most farmers are also not earning living wages and depend on a family member with an off-farm job to provide health insurance and retirement benefits. A 2004 study of organic farmers in Wisconsin found that on the smallest farms—where much of the work is done by hand—the farmers were earning about $5 an hour. On the larger, more mechanized farms, the farmers earned up to $11.36 an hour. As one extension agent has related to me, many of the farmers he works with earn less than $10 an hour. Highly skilled and experienced organic farmers, after 20 years of building their craft and their business, are earning about $20 an hour. This is a fraction of the wages other careers could have provided them.

With the investments needed to start a farm, new farmers are unlikely to make real, living wages for five to ten years.

As a result the Agricultural Justice Project (AJP)[1] began seeking an effective way to bring attention back to the social and economic needs of the farmers and farmworkers who practice organic and sustainable agriculture. AJP has drafted social stewardship standards, translating the abstract notion of social justice into the concrete terms of pricing for farm products and working conditions on farms. The project functions on the idea that mutual respect and cooperation among the people who grow and sell food will result in a triple win for farmers, food workers and the people who eat that food. In 2013, lentils and grains from the Farmer Direct Coop in Saskatchewan began appearing with AJP's Food Justice Certified label in coops in the Pacific Northwest. In 2014, the first farms and food businesses in California and New York have completed AJP certification.

It's time to take the abstract term "local food system" and turn it into a living network, a community of producers and their co-producers (a term coined by Carlo Petrini, founder of Slow Food). Co-producers are consumers who understand the realities of farming and act accordingly. Not only do they eat the fresh and delicious products of local farms, they also willingly seek out local producers and pay them prices that support their farming. Through the practice of exchanging money for food, producers and co-producers can take responsibility and control our destiny.

Together, we can share the whole loaf—food that is Local, Organic, AND Fair!

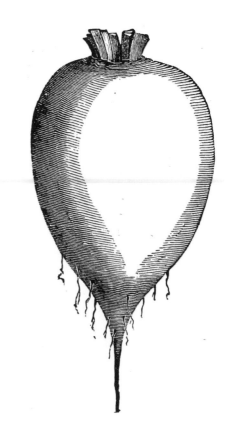

---

1   You can read about AJP standards, their history, and policies at www.agriculturaljusticeproject.org and about the Domestic Fair Trade Association and its evaluation process for fair trade claims at www.thedfta.org and www.organic-center.org

Abraham Lincoln Said . . .

As I would not be slave, so I would not be master. This expresses my idea of de m o c r a c y— Whatever d i f f ers from this, to the extent of the difference, is no democracy.

‡ ‡ ‡

This country, with its institut i o n s, belongs to the people who inhabit it. Whenever they shall grow weary of the existing government, they can exercise their constitutional right of amending it, or their revolutionary right, to dismember or overthrow it.

‡ ‡ ‡

If any man tells you he loves America, yet hates Labor, he is a liar. If any man tells you he trusts America, yet fears Labor, he is a fool.

G o vernment is a combination of the people of a country to effect certain objects by joint effort. The best framed and best administered governments are necessarily expensive.

‡ ‡ ‡

Advancement — improvement in condition — is the order in a society of equals. As labor is the common burden of our race, so the effect of some to shift their share of the burden onto t h e shoulders of others is the great durable curse of t h e race.

‡ ‡ ‡

When you begin qualifying freedom, watch out for the consequences to you.

EDITOR'S NOTE Interested in following in Lincoln's footsteps and becoming a lawyer without paying the exorbitant prices of Law School? There are options for you, check out www.likelincoln.org (www.theselc.org)

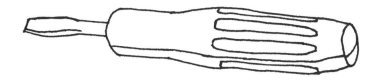

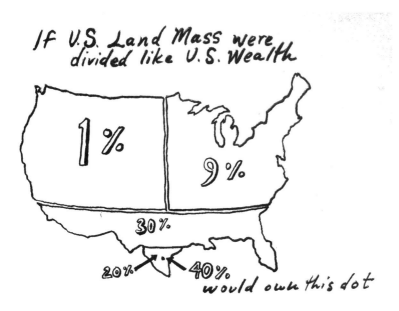

If U.S. Land Mass were divided like U.S. Wealth

1%

9%

30%

20% 40%

would own this dot

MORE TRUTH THAN POETRY
Farming in 1870
The farmer's at the plow
The wife milking the cow
The boys threshing in the barn
The daughters spinning yarn
All happy to a charm.

Farming in 1915
The farmer's gone to see a show
The daughter's at the piano
The madam's gaily dressed in satin
All the boys 're learning Latin
And a mortgage on the farm.
Farming in 1941
The farmer's tinkering with the tractor
Daughter's being made beautiful by Max Factor
Mother's busy cooking from a can
Sonny's at the tavern discussing Sally Rand
No wonder Papa's renting land
And soon he'll be a hired man.

I am not among those who fear the people. They, and not the rich are our dependence for continued freedom- And to preserve their independence we must not let our rulers load them down with perpetual debt.—Thomas Jeferson.

Abraham Lincoln in his message to congress in 1861 said; "Labor is prior to and independent of capital—Capital is only the fruit of labor, and never could have existed, if labor had not first existed.—Labor is much the superior, and deserves much the higher consideration."

THE FEUDAL WORLD

1. THE FEUDAL WORLD

FEUDALISM'S LONG REIGN

With the disintegration of the Roman Empire in the West, beginning with the fifth century after Christ, a kind of paralysis spread over the face of the European continent. Once a unity, as the result of the operations of a hundred and one different economic and political forces, the Western World was now being shattered into a thousand fragments. Localism began to replace Roman imperialism. In place of a far-flung commerce on the basis of which goods and services had been widely exchanged, a meager agricultural self-sufficiency began to appear. In place of the pomp and circumstance, at least of the upper classes, there began to emerge a universal misery and squalor. Insecurity, hunger and dreadful pestilences held the dwindling populations of Europe in their grip. Mankind's inability to reconcile the many contradictions of the classical civilization had led to its collapse.

For a long time — from five hundred to a thousand years — this local organization of chaos which we call feudalism held the stage on a great part of the European continent. For almost another five hundred years mankind was concerned with the dismantling of the feudal institutions. In England in the seventeenth century and in France in the eighteenth century revolutions were required to smash the power of feudal authority. Only within recent years has the Western World become fully free of the restraints and controls of feudalism. And it is a sign of our modern disorganization that Fascism in Italy and Germany has been seeking to reimprison men within the contracting walls of the feudal society and economy.

FEUDALISM BASED ON AUTHORITY

Feudalism was a society based upon status. The individual was not free to come and go as he pleased. Nor was he free to raise himself from a lower to a superior station. The serf was attached to the soil just as his kine were; and when the manor changed hands so did he. Feudalism, too, was a natural economy as opposed to the money economy of capitalism. Men produced agricultural goods for their own immediate use; town life languished; what exchange took place was not far from a crude form of barter.

Over the whole system hovered the authority of the Catholic Church. By preaching the doctrine of salvation through sacraments and good works it lulled the minds of men into an unhesitating acceptance of the *status quo*. The rewards for a blameless life (fealty to one's lord, devotion to the Church) were to come in the hereafter. In this vale of tears men were to bear the burdens of class oppression and misery and sudden death without complaint.

Are Dynasties of Money

There is an old American tradition that "it is three generations from shirtsleeves to shirtsleeves," or "from overalls to overalls." That may have been true at one time, but reports recently issued by the O'Mahoney Monopoly Committee reveal that those who own most of what is worthwhile in America have discovered a way to beat the tradition.

Not so long ago, LABOR summarized a report prepared for the O'Mahoney committee by the Securities and Exchange Commission. It showed that 13 of the super-rich families control the 200 largest "non-financial" corporations in this country. According to the "shirtsleeve" tradition, that might suggest that wealth and control were in the hands of the "first generation." But that is not the case. According to the report:

Dynasties Established

"The Mellon group is now in the third generation, while the Rockefeller and duPont groups are in the second and third generations.

"Most of the other family groups are also in the second and third generations; for instance, the Duke Hartford, Widener, Harkness and Woolworth holdings. Few of the large stockholdings are still owned by the founders."

Henry Ford dominates his automobile empire, but the second generation of this "royal family" is already sharing control and the third is coming up.

The S. E. C. report tells how these families perpetuate their wealth and power, regardless of the ability of the sons and grandsons.

Dead Hands Hold Reins

"Family holding companies, trusts and foundations have made it possible to keep control centralized in the hands of a few persons," the report says.

"An important part of this centralization is played by the appointment of the same trustees for a large number of trusts, having different beneficiaries.

"Thus, practically all the trust funds set up within the Rockefeller family are administered by the Chase National Bank, which is under Rockefeller control.

"Most of the Mellon family trusts are administered by the Union Trust of Pittsburgh, controlled by that family. The duPonts use the Wilmington Trust Company and the Deleware Trust Company, which they own."

Doesn't Trust Descendants

To put it simply, "papa" acquires great wealth. If he puts that wealth directly in the hands of his sons and grandsons they will probably lose it and have to go to work.

So, with the aid of clever lawyers, he forms family holding companies, trusts and foundations, managed by hired investment experts, and men skilled in advising how to dodge income and inheritance taxes.

Thus the wealth and power are perpetuated; "papa's" descendants are saved from the necessity of working for a living. They do not even have to clip bond coupons. This is done for them by a hired expert.

The interesting facts revealed by the S. E. C. report constitute something new in the history of America, because family holding companies, trusts and foundations are comparatively recent inventions.
—From the weekly newspaper, Labor.

Tenant Farmers in Each State.

The figures show how many farm families are tenants out of every 100 families living on farms in each state. The other farm families owned the farms upon which they lived. The census of 1890 showed an average of 34 tenant farmers in the whole United States out of every 100.

Ownership of Farms in Each State.

The figures show how many farms were mortgaged out of every 100 farms occupied by their owners. All the other farms occupied by their owners are free of debt. The census of 1890 showed an average for the U. S. of 28 mortgaged farms out of every 100 farms occupied by their owners.

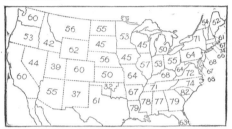

Tenant Homes in Each State.

The figures show how many families occupying homes (not on farms) are tenants out of every 100 families living in homes, averaging 63 for the whole United States. The other home families own the homes in which they live.

Germans are the most numerous of foreigners in every section except in New England, where the Irish lead in numbers.

If the earth were not enveloped with atmosphere, the temperature on the surface would be about 330 degrees Fah below zero.

'Tis a great journey to the world's end.

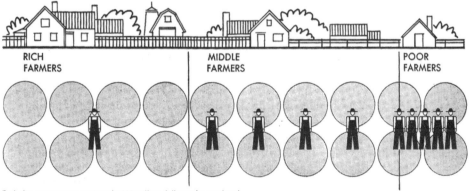

Ownership of Homes in Each State.

The figures show how many homes (other than on farms) were mortgaged out of every 100 homes occupied by their owners. These averaged 28 for the whole U. S., the other 72 % of homes occupied by their owners were free of debt in 1890.

During ten years previous to 1890 Chicago had doubled its population and from the fourth city in rank has become the second. During the same period Omaha increased fourfold, Minneapolis and St. Paul together increased threefold and Denver grew at the same rate, while Kansas City more than doubled its population.

Justice is the rightful sovereign of the world.

CLASS COMPOSITION OF AMERICAN FARMERS

RICH FARMERS   MIDDLE FARMERS   POOR FARMERS

Each disc represents approximately 600 million dollars value produced
Each farmer represents 600,000 farmers
Data for 1929

The Refunding of Farm Mortgages.

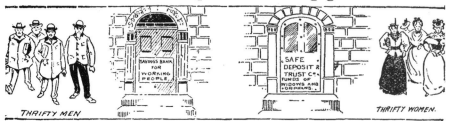

THRIFTY MEN

THRIFTY WOMEN.

The People and Institutions That Furnish the Money Which is Loaned on Farm Real Estate.

RATES OF INTEREST AT HOME AND ABROAD ON REAL ESTATE MORTGAGES.

These are the average rates that prevailed in 1890, according to the federal census, but usurious interest is often exacted.

GROUPS OF STATES (U. S.).	North Atlantic	South Atlantic	North Central	South Central	Western	United States
For all mortgages on real estate	5.50	6.66	7.43	8.08	8.89	6.60
" " " farms and homes,	5.53	6.45	7.33	8.02	8.96	6.65
For mortgages on acres,	5.76	7.02	7.62	8.44	8.93	7.36
" " " farms,	5.62	6.64	7.43	8.05	9.08	7.07
" " " lots,	5.51	6.36	7.17	7.40	8.84	6.16
" " " homes,	5.48	6.26	7.12	7.96	8.81	6.23

	Per cent.		Per cent.			Per cent.
Africa,	6 to 15	Belgium,	4 to 5			
British Asia,	6 " 9	Denmark,	4	Russia,	6½ to 7½
China,	7 " 30	France,	4 " 6	Sweden,	6
Japan,	10 " 15	Germany,	3½ " 5	Great Britain,	3 " 5
Turkey and Asia,	9 " 50	Greece,	6 " 10	Canada,	5 " 8
Central America,	9 " 12	Italy,	4½ " 9	Mexico,	7 " 18
Australasia,	68	Netherlands,	4 " 4½	Argentina,	7 " 18
Austria-Hungary,	4 " 5	Portugal,	4½ " 6½	Brazil,	6 " 12

The Farmer Who Borrows the Money

And the effect upon him of refunding, at reasonable rates of interest. His present burden of $70,000,000 interest per year, the rate being over 7 per cent, almost crushes the life out of him. Reduce it to 6 per cent, and he can get up, at 5 per cent it is easier, at 4 per cent he can lug it readily, while 3 per cent, or $30,000,000, he can walk off without effort.

Mortgages on real estate in the United States are this year being paid off more rapidly than for several years. In most foreign countries loans on real estate are increasing in number and amount much more rapidly than in the United States. The proportion of mortgaged farms or homes in the United States is comparatively small. But the mortgaged farms of America, carrying as they do an average debt of only about

Lost time never returns.

Ownership, Labor & Justice

Jonathan Magee

"When are you going to start your own farm?"

For young farmers, the issue of farm ownership is a loaded one. We hear that more of us are needed to take over from retiring farmers, and many of us long for the independence and status that ownership confers. In the current back-to-the-land movement, we have an informal map toward ownership: from humble beginnings as an apprentice, to middle management, to full management, and finally to ownership. By this route we supposedly become established and achieve independence, the way our forebears did before us.

This career path hearkens back to the 'agricultural ladder,' a framework that gained currency around 1920 and shaped debates on farm tenure until the 1950s. First proposed by R.T. Ely, the ladder laid out a bright future for the hard-working farmer, rising naturally from unpaid family labor to wage work as a field-hand, to tenancy, and finally to full ownership

of a farm enterprise. This career path lined up perfectly with the Jeffersonian Creed, which honored the free and independent yeoman farmer as the ideal citizen. According to the ladder scheme, if a laborer was diligent and industrious, he would attain that final, noble state.

Today many non-profit and governmental organizations are once again devoting significant attention to connecting younger farmers with ownership opportunities, to ease the looming retirement of half of America's farmers. With that effort in mind, we would do well to revisit the agricultural ladder theory as it was laid out in the last century. The model was a resounding failure—both in its attempt to describe tenure patterns and in its prescriptions for policy.

Looking back at the period since 1900, "hard work and frugality on the part of the rural landless in the United States did not, by and large, result in upward tenure mobility" (Kloppenburg & Geisler).

Despite our national affection for personal advancement, there was little opportunity for non-owners to improve their lot: workers faced an entrenched class of landowners, low wages, limited access to capital, and a market and government hostile to the small-scale farm. Yet contrary to the mounting evidence that class mobility was barely achievable at best, the idea of the ladder persisted. For, as Kloppenburg and Geisler point out, "the ladder extended the life of the Jeffersonian creed which exalted agrarian virtue and dispersed private property as the underpinnings of democracy…in a period when agrarian dissatisfaction was turning increasingly radical."

The effect of the ladder was to pacify workers and drive them to greater diligence, while implying the inferiority or laziness of those workers who couldn't reach the ladder's top rung. The class structure of rural America had become rigid, but to admit as much seemed to be admitting the failure of our ideals. At a time when many farmers were being driven out of agriculture for lack of livelihood, the agricultural ladder served as a smokescreen for an agrarian transformation, a daydream for bureaucrats and commentators, and an empty promise to generations of rural people.

Many of us who have moved through the ranks of farmworkers and apprentices are still aware of this silent class struggle, and even today it feels terribly impolite to speak of it. This has been a learning experience for those of us from urban, middle-class backgrounds. Part of the excitement behind the back-to-the-land movement is the thrill of jumping between worlds: if we can take an airplane between continents, surely we can move out of the city and become a farmer, right?

This presumed mobility speaks to the privilege of our position—though some of us may move in circles where college degrees, social advancement, and a family safety net are commonplace, most people in the world do not have these advantages. The reality of class is more complicated, and most of the world's people have limited mobility and options for advancement. To paraphrase Alessandro Portelli, the more that our society assumes total mobility, the less we are able to see the real barriers that exist.

Even today the promise of the ladder suggests endless opportunity, but experience suggests otherwise. Few people in the back-to-the-land movement have the resources or connections to establish their own farms, while the rest struggle to find steady work. This is not for lack of effort on the part of the newcomers.

EDITORS NOTE "The Farmers' Holiday" was a radical movement to withhold farm products from the market, organized to resist low prices which did not allow farmers to meet mortgage and tax payments.

Neither is life so simple for those farmers still-on-the-land. Some landowners and businesspeople have weathered the storms of foreclosure, consolidation, and encroaching development and come out prosperous, while contract farmers labor in peonage to distributors and meatpackers, and hosts of marginalized farmworkers escape notice. Retirees, landlords, developers, and investment funds all buy up farmland to rent or hold. We cannot forget those no-longer-on-the-land, the many farmers, tenants, workers, and their families who lost their livelihoods, quietly left their homes, and sought refuge in cities. This last group have been excluded from debates around agriculture just as they've been excluded from farming itself. Many millions of people have been forced out of the countryside since the founding of our nation, and their unexplained absence points to a gaping hole in our understanding of history.

Unsurprisingly most public commentators failed to account for those who were not climbing upward. The ladder model was just another formulation of an ideology with much deeper roots. Thomas Jefferson was one in a long line of thinkers extolling private property and personal industry, stretching back to Martin Luther and John Locke. These intellectuals—and they are legion—left an indelible mark on modern religion, economics, law, and government. It's little surprise that talk of personal success and property slips easily into moralizing and judgment.

During the 1930s tenant crisis, for example, government committees concluded that farm ownership, as opposed to renting or wage labor, was "definitely related to character and patriotism. It is conducive to high character and good citizenship" (Grubbs). In this commonly-held view, ownership and economic success are signs of virtue, while poverty and failure are shameful by implication. This ideology has often been taken to its extreme by politicians and commentators who felt "that poverty was the result of shiftlessness and incompetence," who had "no confidence at all in any scheme to cure these faults of character" (Tugwell).

Claims of "character and patriotism," when uttered by politicians, are often mere pandering to moneyed and propertied interests. However, these ideas are not voiced strictly by our public servants but feature prominently in conversations about agriculture across the political spectrum. For those of us who wish to rebuild agriculture in a more just and equitable form, we must recognize the ideology hidden in the way we farm. I propose a few questions to guide our efforts: First, what does it mean to own a farm today, and what has it meant historically? Secondly, if we as a nation have always prized the family farm, what have these agrarian values meant in practice? In other words, is there a gap between rhetoric and reality when we voice our love for the family farm? Present food movements often promote smaller farms as the key to a more

ecological agriculture, but are small farms more economically just?

To ask these questions is not simply to doubt certain core tenets of our American identity or our understanding of history. By recognizing the complexity of ownership, we also gain a vision of radical possibilities—the sort of possibilities which can bridge the gap between food justice as a dream and as a reality. The idea of owning one's own farm may appeal on many levels, but the pursuit of individual ownership will not resolve the fundamental inequality that has plagued American agriculture. Looking forward, we can build an agrarian future beyond the narrow frame of the past. To do so, we must relieve farmers from exploitation by markets and the state, claiming food—its production and its consumption—as a central rite of community life. At the same time, we must extend solidarity to all farm people, owners or not. In this way we can fulfill the radical potential of present food movements: farming as a practice of liberation.

To ask these questions is not simply to doubt certain core tenets of our American identity or our understanding of history. By recognizing the complexity of ownership, we also gain a vision of radical possibilities—the sort of possibilities which can bridge the gap between food justice as a dream and as a reality. The idea of owning one's own farm may appeal on many levels, but the pursuit of individual ownership will not resolve the fundamental inequality that has plagued American agriculture. Looking forward, we can build an agrarian future beyond the narrow frame of the past. To do so, we must relieve farmers from exploitation by markets and the state, claiming food—its production and its consumption—as a central rite of community life. At the same time, we must extend solidarity to all farm people, owners or not. In this way we can fulfill the radical potential of present food movements: farming as a practice of liberation.

Selected Bibliography & Recommended Reading

Alkon, Alison Hope. *Black, White, and Green: Farmers Markets, Race, and the Green Economy.* Athens: University of Georgia Press, 2012.

Davis, John Emmeus. "Capitalist Agricultural Development and the Exploitation of the Propertied Laborer." in Frederick H Buttel and Howard Newby, *The Rural Sociology of the Advanced Societies: Critical Perspectives.* Montclair, N.J.: Allanheld, Osmun, 1980.

Grubbs, Donald H. *Cry from the Cotton: the Southern Tenant Farmers' Union and the New Deal.* Chapel Hill: University of North Carolina Press, 1971.

Guthman, Julie. *Agrarian Dreams: the Paradox of Organic Farming in California.* Berkeley: University of California Press, 2004.

Kloppenburg, Jr., Jack R. and Charles C. Geisler."The Agricultural Ladder: Agrarian Ideology and the Changing Structure of U.S. Agriculture," *Journal of Rural Studies* 1:1, 59-72.

Post, Charles. "The Agrarian Origins of US Capitalism: The Transformation of the Northern Countryside Before the Civil War," *The Journal of Peasant Studies* 22:3, 389-445.

Going Home

Douglass DeCandia

the less i choose to carry
the easier will be my travels

and wherever I settle
there will be a place for everything.

Bearberry Honeysuckle

Reflecting on
Black Farmers In America: 1865–2000

Jean Willoughby & Sam Hyson

In the wake of the Civil War, ambitious land reform plans were quickly abandoned, and most black farmers ended up as tenants or sharecroppers. Loans for farm supplies secured with crop liens put many farmers, both black and white, into a persistent state of debt. Whereas tenant farmers rented land, provided their own tools and draft animals, and received a portion of the harvest, sharecroppers rarely owned their implements and depended on landowners both to sell the harvested crops and pay down their accounts. As early as the 1870s, black farmers banded together informally to withstand pressures to sell their crops cheaply.

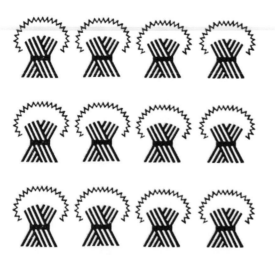

RECONSTRUCTION AND EARLY 20th CENTURY

Between 1886 and 1932, several types of initiatives encouraged black independent farming:

Education
The work of Booker T. Washington and the Second Morrill Act of 1890 increased black farmers' access to agricultural education.

Cooperatives and unions
In the 1880s and 90s, the Farmers Alliance was an important cooperative movement. A branch called the Colored Farmers' National Alliance and Cooperative Union (CFNACU) was started in 1886. The CFNACU helped members pay land mortgages, and in 1891, organized a general cotton harvest strike. Robert Lloyd Smith organized the Farmers' Improvement Society of Texas (FIST) in 1890. FIST was a movement for black cooperatives, and by 1900 had 86 branches in Texas, Oklahoma, and Arkansas. Outside the Alliance and FIST, informal cooperatives and church associations predominated, because formal black cooperatives were not tolerated.

Land Purchase and Resale Projects
Booker T. Washington helped raise capital for several land purchasing and settlement projects, including the Southern Improvement Company (SIC), started in 1901. SIC purchased and subdivided a 4,000-

acre tract of land, and the Tuskegee Farm and Improvement Company (Baldwin Farms) did the same with an 1,800-acre tract purchased in 1914. These projects did not achieve long-term success partly because they were too dependent on the volatile cotton market.

Farm Diversification and Self-Sufficiency

Booker T. Washington influenced the development of the black extension agent service, led by his former student Thomas Campbell. Black extension agents promoted home gardening, crop diversification, and food preservation to make black farmers more self-sufficient. This strategy had both an immediate and

lasting impact.

THE GREAT DEPRESSION AND THE NEW DEAL

The onset of the Depression in 1933 led to a more interventionist role of government in agriculture. The Federal Farm Board was established in 1929 to coordinate agricultural organizations, including cooperatives. The Agricultural Adjustment Act (AAA) of 1933 controlled crop yields for several commodity crops. In order to increase cotton prices, the AAA reduced farmers' production, in turn displacing black and white tenants and sharecroppers. Black farmers were often excluded from land ownership programs within the AAA. However, some displaced sharecroppers and tenants were aided by programs of the Resettlement

Administration and the Farm Security Administration (FSA). These programs lent money for the purchase of farmland and machinery, and also promoted farming cooperatives, including land-lease cooperatives, which leased, subdivided, and subleased plantation land to family farmers. The FSA also implemented land purchasing programs similar to Booker T. Washington's.

FARMS AND FARM LAND

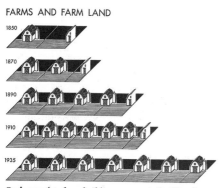

1850

1870

1890

1910

1935

Each complete farm building represents 1 million farms
Each complete land division represents 160 million acres

(Solid areas—unimproved, shaded areas—improved)

THE CIVIL RIGHTS MOVEMENT

Interracial tension during the Civil Rights Movement prompted many black farmers to form cooperatives. Some cooperatives formed in response to discrimination and boycotts against farmers who were NAACP members. One of the most politically active and widely publicized black cooperatives formed in the late 1960s was the South West Alabama Farmers Cooperative Association (SWAFCA). The Southern Cooperative Development Program (SCDP) and the Federation of Southern Cooperatives (FSC) were founded in 1967 to assist farming cooperatives and provide a wide range of services to Southern farmers. The Emergency Land Fund (ELF) was organized in 1971 to assist black farmers in Mississippi and Alabama with land retention. In 1977, ELF received federal funding to establish land cooperatives that functioned similarly to credit unions. In 1985, FSC and ELF merged as FSC/LAF (Land Assistance Fund), which continues to establish cooperatives.

RECENT DEVELOPMENTS

In the early 2000s, FSC/LAF included approximately 75 cooperatives and credit unions. Most of the Southern cooperatives had fewer than 30 members. Many black farmer cooperatives are moving towards producing high-value crops and value-added products.

A major challenge of cooperatives is to increase profitability enough to sustain their growth and attract new members. Consolidation and regional organization of cooperatives may be one effective strategy to increase profitability.

Lastly, the USDA's Rural Business-Cooperative Services assists farmers involved in cooperatives and has established dozens of cooperative development assistance centers. Black farmers have developed creative and enduring responses to the challenges of discrimination and disinvestment within US agriculture. Too many black farmer initiatives have been hampered by a hostile social and political climate that devalued and discouraged their efforts. Black farmers have demonstrated powerful examples of community-based economics, cooperative business initiatives, and collective land ownership arrangements. These innovative approaches should be recognized, documented, and studied by all who would hope to create a more just and sustainable agriculture in the US. We still have much to learn from the history of black farmers in America.

Bibliography

Reynolds, Bruce. J. 2002. *Black Farmers in America, 1865-2000: The Pursuit of Independent Farming and the Role of Cooperatives.* USDA Rural-Business Cooperative Service.

AN 18th CENTURY VIRGINIA ESTATE

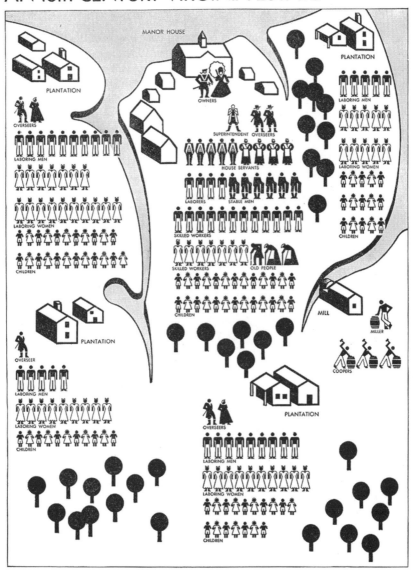

March

Opportunity

Chapter 3

Look to Maine

Maranda Cook

They say that "As Maine Goes, So Goes the Nation." Our motto is—Dirigo—"I lead." And for centuries on the economic front, this has been true.

We've exported our natural resources whole-heartedly and lost our industries to global outsource. We've sent our young folks far from home—until recently this was in search of promise, but now it's only out of habit. We've ignored poverty and racism, had our share of indecent affairs with corporate developers, pretended bottled water is great, and taken money from Washington to enact bad policies at home.

As we've gone, so has the nation.

It's only right that I share with you where we are going next, as it pertains to food and agriculture. And it's only right to disclose that I run three Maine food businesses, Crown O'Maine Organic Cooperative (distribution), Northern Girl, LLC (processing), and Fiddler's Green Milling Company (value added & mailorder). I will add (with your patience appreciated) that the following are my opinions and no one else's unless they chose to share them.

Maine has legions of young farmers and food producers. We have distribution, co-packing, feed, processing, slaughtering, winter farmers markets, and CSAs galore. We have collectives and coops and ninth-generation family farms. We include 4 sovereign Native American tribes, some with farms.

The public has access to Maine's Legislature, and we use it. We saw this recently in the passage of GMO Food Labeling Legislation—it's coming, folks. One reason we won is because Monsanto didn't realize in time that Mainers, conservative and liberal, are 94% in support of labeling. Do not lose hope. The little places matter.

But Maine is headed for a dangerous period. Local food production thrives under the steam of passion and youth, yet incomes of many Mainers are generally inadequate for subsistence. The issue of affordability of food beats around the bush of family poverty.

Maine has not lost its ability to feed our own, and resilience is not as arbitrary a term as it may read. Meat can still be shot off the back porch (depending on your

there. They are said to inherit the earth, so pay attention to where they buy property.

Maine has fish and seaweed and salt. All of these are rather less contaminated than in other parts of the world if you are into that kind of thing. Our local food movement here is both aquatic and terrestrial. You will see Mainers' health change radically as the localization of both these food systems meets the health care directives coming along.

Maine has an aging population, including farmers. With incomes low and aging high, we need young people to come and support our communities, our businesses, and our farms. We need young people to bring business opportunity with them, and to improve those family wages I mentioned before. We need to increase Maine families' capabilities to buy Maine food rather than sending ourselves back into the cycle of depressing farm prices

Maine is a creative, beautiful, unusual place to call home. I am its daughter to the core. I invite you, dear reader, to visit the northernmost organic farm in New England— my family homestead we call the 'Agricultural Embassy of Aroostook'. Stop by for a weekend at our farmstead in Grand Isle, Maine.

zip code I suppose). We have a Hunters for Hunger Program. I have butchered a (freshly) road-killed deer at a time when I would have qualified for food stamps. I never realized I needed them, but no Maine family should have to live in hunger.

Maine has vast acres to the north available for farming ventures. It is cold there, but it also rains. The land is affordable. Scores of young Amish families are moving

Closest major airport is Quebec City, but you can fly from Boston to Presque Isle for reasonable time and money too.

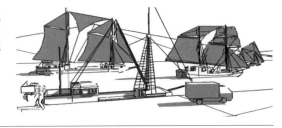

"Though a peasant's life is not a fat one, it is a long one. We shall never grow rich, but we shall always have enough to eat."
—Leo Tolstoy, *How much land does a man need?* (1886)

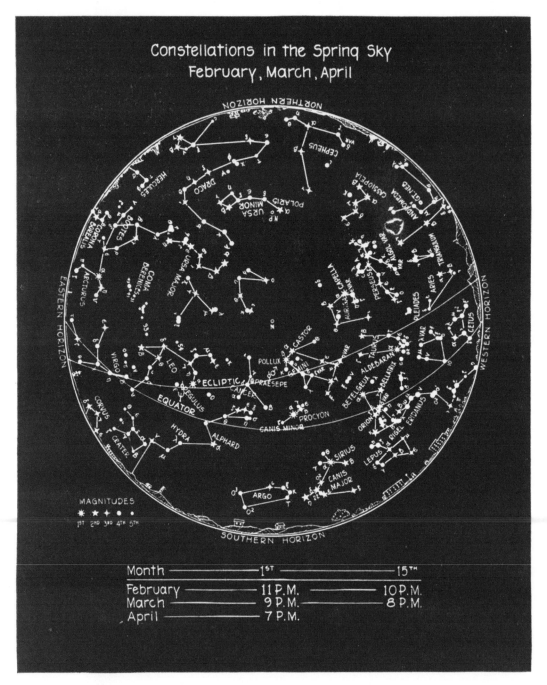

Constellations in the Spring Sky
February, March, April

Month	1ST	15TH
February	11 P.M.	10 P.M.
March	9 P.M.	8 P.M.
April	7 P.M.	

GENERALLY speaking, the stars rise about an hour earlier every two weeks, so that a chart which represents the sky at 9.00 P.M. on one night will represent it at 8.00 P.M. two weeks later, and 7.00 P.M. a month later.

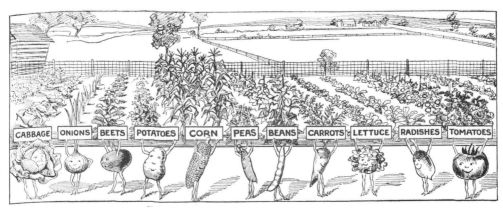

CABBAGE ONIONS BEETS POTATOES CORN PEAS BEANS CARROTS LETTUCE RADISHES TOMATOES

Greenhorns Romance

Ali Palm

It was late winter in southern New England a couple of years ago. As an aspiring farmer who had somehow ended up in college (but was still living at home and keeping up with the milking), I was spending time with people whose lives were going to be very different than mine. I was starved for friends with agriculture on the mind, and I really wouldn't have minded a boyfriend either.

The NOFA winter conference was full of workshops, including one for young farmers courtesy of the Greenhorns, and a young farmer from Connecticut. I slipped in, hoping for a glimpse of these other young farmers in megalopolis. The room was full of middle-aged women (who asserted they were young at farming, which is why they came), a couple grad students who needed some 'insight on agriculture' for a class, and a ripped young man sitting on the far side of the room. His family ran a nursery. The workshop was good—hearing about the Greenhorns, reminding myself that there were other young people out there farming, and remembering that the bottom line isn't just about cash—but afterwards the guy left before we got a chance to talk. Or perhaps, depending on your interpretation, I couldn't get up the guts to go talk to him.

There was another workshop slot, and then I found my mother. We got ready to go home and I hadn't seen the guy again. I was holding a giant bag of apples we'd bought and waiting, when he came along. He said that I'd mentioned goats during the workshop, and so we talked about how he was thinking about getting goats, and what it's like having goats. And hey, my family was possibly looking for someone to goat-sit at some point the following summer... so he went home with my phone number. We ended up dating that summer, and he never did goat-sit. But the lesson was thoroughly learned: don't pass up an opportunity to connect with the people you live like, especially when they're few and far between.

Illustration by Julia McDonald

Of Carob and Acorn

Adam Huggins

On a recent family visit to San Jose I found myself walking down a familiar suburban street on a sunny November day. There, amongst privets and peppertrees, I was suddenly accosted by an ample, brown pea-shaped pod with the distinct odor of Limburger cheese. Intrigued, I looked up and found the offending tree to be a vaguely exotic mess of dark brown bark and thick, shiny opposing leaves. I ran home to consult my tree books and discovered that I had been standing under the canopy of the Mediterranean Carob, *Ceratonia siliqua*. What could such an illustrious and mysterious dynamo of a tree be doing, acting the part of a well-behaved, unassuming ornamental in front of the supermarket on my street?

At one time not so long ago, the fertile valley bound by the Santa Cruz and Diablo ranges on the southern washes of the San Francisco Bay was known as "The Valley of Heart's Delight." It was a worldwide center for fruit production, only recently wrested from the native Ohlone people who lived comfortably on its crops of acorn mast, wildflower seed, and game. A brief century after its initial colonization it underwent a second, equally sudden conquest: today it is known as "Silicon Valley," and only relics of its past have survived.

THE TOLERANT TREE WINS. HOW?

I returned to the tree and collected a tidy harvest of pods, which tasted sweet right out of hand. Back in the kitchen, I boiled them for a half-hour to loosen them up then cut each pod in half and separated out the handsome black seeds. The pods I cut into chunks and set on screens. Dried, they made a unique addition to trail mix. Powdered, they became a baker's delight. Carob was an instant culinary staple, blending well with the fall crop of persimmons foraged off neglected suburban trees. Further research has unearthed many interesting tidbits. The carob tree has been cultivated for ages by farmers in the Mediterranean on the driest land with the steepest slopes. The seeds, which seem to be of relatively uniform weight, were the original "carat" measurement for precious stones and metals. They are also processed commercially for carob bean gum (CBG), which is used in everything from ice cream to flocculating fluid. The pods, which are thought to be the proverbial "locust" that St. John the Baptist subsisted on in the desert, are high in sugar and protein and make a valuable food for both humans and animals. And while the dieting fads of the 1980s proved that carob is not a chocolate substitute, it is nevertheless delicious on its own merits in raw vegan fudge or mouth-melting coffee cake.

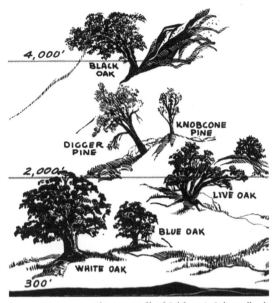

Elevation profile of California's Oak woodlands

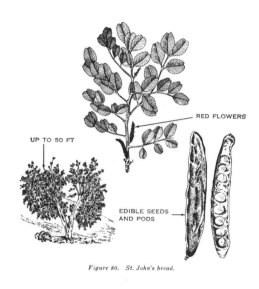

Figure 80. St. John's bread.

Carob Tree, or St. John's Bread

Most importantly, as J. Russell Smith points out in Tree Crops, the mature carob is a dependable, heavy-bearing, shade-casting tree that is perfectly adapted for hot, arid climates and poor hillside land.

Central and Southern California consist of overpopulated, heavily urbanized valleys; overgrazed, barren hillsides carpeted by non-native annual grasses and studded with the remains of vast oak forests; and destructive irrigation-dependent agriculture squeezed in between. The demand for water is overwhelming, the water infrastructure hugely complex and surprisingly vulnerable, and many ecological signs point firmly towards desertification. And the oaks on those hillsides look mighty lonely...

In light of this, perhaps the most amazing aspect of the carob tree in this country is its stupendous obscurity. It seems like an excellent choice for culture in California on any scale. Some brave folks have tried. During the Great Depression, carob trees were planted around Los Angeles schools to provide free, nutritious snacks for school children. In the 1950s, J. Elliot Coit assembled

a modest carob orchard near Vista, CA, which was subsequently destroyed (in fact, none of the few extant commercial plantings remains). Only a single nursery (PapayaTree, in Granada Hills) offers a named variety (Santa Fe) in the whole of the state. Carob in California is almost entirely the province of us urban foragers.

As I write, the first fall rains are falling on the little homestead where I live in the Klammath Mountains after many dry months of summer, the natural season of dormancy in California. I rejoice in the mists hanging over the conifer-clad hills and the sweet earthy smell of relief and rejuvenation rising out of the orchards. But I cannot help thinking of the hills above my family home in the Bay Area. In my mind I can see hundreds of carobs planted amongst the live oaks on those hostile slopes, providing sustenance and shade to the denizens of the valleys, playing a supporting role in a new reconciliation ecology. In the meantime, as my carob supplies dwindle and fall rolls around, I am plotting my next trip down to my favorite street-corner supermarket—*Ceratonia siliqua*.

Give the Liars a Chance.

Wanted—A legion of able-bodied liars and slanderers, to lie about and slander the Union Labor party and all connected with it. All applicants must be at least 21 years old, and a voter in one of the *"grand old parties,"* and windy enough to howl *"anarchist" "crazy green backers," "cranks"* etc., for 10 hours each day, and hiss like a reptile, or bray like an ass every time they happen to see a real live man. It is not required or expected that applicants have either intelligence or brains, as the 5 by 7 runts of the "damphool" variety, now posing as editors of the g. o. p. county papers will answer as substitutes where *men* are not wanted.

Members will not be expected to know what they are talking about, or to have any ideas of their own, but they must be sufficiently obedient to obey their bosses, vote the straight ticket, and lie like the very devil about everything they do not understand! As a sample of the kind of material wanted, please examine the *Editor* of the *Republican Record* of Erie, Neosho Co., Kan.

It is proposed to organize a donkey brigade, composed of the most narrow and malicious liars to be found in the two old parties and pay them in free whiskey and red bandannas.

Old party editors of little county papers, who have been born in ignorance, nursed on hypocrisy, and fed on political garbage until they are stunted in brains, starved in body and warped in morals, will be made corporals for distinguishing themselves as heroic slanderers or strategetic liars, and we would advise them not to miss this chance, as it is the only opportunity they will ever have of doing anything worthy of notice. You g. o. p. time-servers whose miserable itch for scribbling has deluded you into the notion that your pens are mightier than pop-guns, quit your frantic efforts to imitate *manhood* and let your tongues wag in the more congenial trade of lying! Lie, and lie furiously and even religiously and aristocratically for the benefit of your starved bodies and the glory of the "grand old party!" Don't try to get a new idea into your heads; there is not room enough for comfort and it might make you dizzy. Stick to downright and bare-faced lying, and take no risks of causing mental paralysis by coming near the truth. Don't masquerade any longer under a lion's skin, for both your ears and voice betray you. Join the Donkey Brigade of Ancient and Incorrigible Liars, and serve your masters in a way congenial to your tastes, and pleasing to your owners. Any one of you who may accidentally tell the truth about the Union Labor party will be considered unworthy to kiss the great toe of your party bosses. If you are an abortion, an excrescence or misfit in nature, be thankful that your masters have more use for such rubbish than they have for men. Just keep right on at your legitimate business of hatching illegitimate falsehoods, for it don't hurt us, and it seems to do you fellows lots of good. When you poor little fellows presume to blubber about "decency" or "blackguardism," you remind me of the story about the Irish prostitute, who, when exposed under the glare of the policeman's lantern, indignantly howled: "Aren't ye ashamed of yerself? Ye blackguard, ye!" How you do squirm and squeal in sham decency and chagrin when we throw any light on the thieving and damnable practices of your masters! Not being able to meet any of the arguments of Union Labor men in a manly way, your only alternative is to crawl into your holes and hiss and wriggle like reptiles. Now then, we want you to lie about us, for every silly, miserable lie hatched out of your empty heads only adds another to the overwhelming load which the two old parties are staggering under, and as lying is about the only talent you have, you had better use it as the only means you will ever have to gain either a livlihood or notoriety.

The time is coming when the American people will say to you as Gratiano said to Shylock: "Beg that thou mayst have leave to hang thyself!"

Carthage, Mo. WATSON HESTON.

Clash Of The Trenchermen

Steve Sprinkel

Horticultural surplus begets depravity, recklessness, competition. Amid plenty, we prove to the Gods that we are existential creatures no wiser or far-sighted than dogs. In Spain, tomato abundance begets La Tomatina, a pitiable wasting frenzy in Bunol, a yawnable speck on the Valencian agricultural landscape, where youths have since 1945 pelted one another into submission with perfectly ripe tomatoes that chilly Poles and Latvians would pay dearly to eat. Milwaukee, Wisconsin; Esparraguera, Chile; Dongguan, China; Reno, Nevada; and Kamtaka, India now throw tomatoes. The Grommus hosted a battle last year at my farm that was enjoyed by everyone except the poor fool who got a concussion. Mr. Ruiz was not a fan of the spectacle. Food is sacred to him.

Neither do I appreciate these violent orgies, but when there is too much watermelon in the field, an eating contest seems like a natural method to get rid of some of the fruit, lest it rot. On the 14th of August, eight brave participants gorged themselves on delicious Crimson Sweet watermelons at the El Roblar Sandia Fest, at the Farmer and the Cook in Meiners Oaks. The total weight of competition fruit consumed may have been a mere fifty pounds, but thirteen times that volume was enjoyed by the forty attendees gathered to cheer on the pitiable contestants. The sidewalk was black with seeds.

Generosity, obviously, is why one thinks of having a "contest." By the way, one should grow copious quantities of watermelon, poised to ripen when the summer begins to burn in order to benefact more easily. The best watermelons are huge. The communal quenching, sublime. There is no more absurd reason for growing 2400 linear feet of watermelon than giving a lot of it away to the unsuspecting. If I produced an overabundance of Tilsit cheese I don't think the recipients would have had such a rapturous response as they did when presented with a firm, free, generous cut of *Citrullus lanatus* (Thunberg). The famed Swedish botanist Carl Thunberg has been tagged for formally classifying the South African native, beloved worldwide, iconic in Georgia and Texas, revered on Hokkaido, where the primo Densuke melons can go for $250 each. We probably grew between eight and ten tons.

The Watermelon Eating Contest was governed by a fifteen minute consumption period. Head Referee Dillon Scheps quickly created Rules and weighed the melons out beforehand in consideration that the true measure of who ate what would be determined by weighing the left-over rinds piled before each miserable contestant. The subtraction, we hoped, of rind from whole melon would yield the actual consumed quantity. Magnificent trenchermen and women buried their faces in red flesh with abandon. Dark Horse Justin Bastien, who boasts a renowned, bottomless appetite, led the front pack of Wiley "Grommus" Connell and Chaz "Get" Everritt. Sarah Nyman and mystery female entrant "Seedless" blasted through their portions, nearly keeping up with the Gentlemen in this sexually indifferent affair.

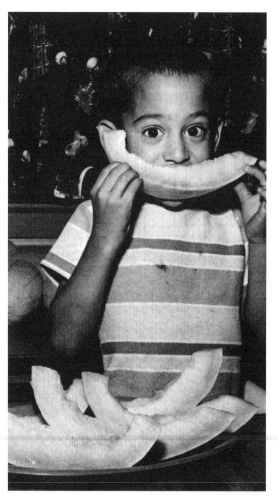

Eric the "Intuitive Grazer" and Jer-Bear, AKA "Grand Master Chai" filled the field. Oddly missing from the contest was John Phaneuf, regionally famous watermelon connoisseur, who begged to remain a field judge because he loved watermelon so dearly he didn't want to take a chance on overdosing on his favorite food.

The contestants, bathed in watermelon sugar, munched and slurped until time was called, a mere nano before unpleasantness erupted. But all held their melon. Now came the judging, but our Referee, perhaps sensing imminent trouble, had precipitously bolted to Refugio Beach for an allegedly pre-planned reunion, leaving behind his hieroglyphic scrawl. These "notes" were intended to guide us in selecting a winner, but it was only a recipe for disaster. It fell to me, who was enjoying a simple, good time giving away thick slabs of red summer fantasy, to don the arbiter's sombrero and take pains to adjudicate.

After I had failed, by awarding first prize to Bastien, then Grommus, by a mere seven tenths of a pound (out of nearly six total pounds consumed per man), claimants to the throne of Watermelon Regent assailed me and I delegated the math to a keen teen who asserted she saw my error (or Scheps'!) and devised a theorem to help us salute the winner. To no avail. The Bear seemed to have eaten much more, but his adversaries ratted him for finger-scraping the melons and leaving a broad pool of juice underneath his seat. The Grazer pled his case so earnestly, and then Bastien raked me mercilessly for switching the protocols from "volume" to "weight" that I tossed it in and awarded 8 Grand Prizes, so all won.

Selected Principles of the Agrarian Trust

COMMONS

Our land is the foundation of society, our economy, and all humanity, it is also home to all eco-systems and wild creatures. Our management and ownership regimes, our notions of private property and water rights do not override the laws of nature. We acknowledge the limits of fragile systems, and that we are land-users in common with wetlands and wild-lands that must be protected from damage.

TRANSPARENCY

Land holdings should be described by clear legal documents that describe the rights of all parties and include accountability and justice. Farmers and landowners both benefit from a written lease agreement with provisions for exit, mediation in case of conflict, a defined process and a strong community of professional farm service providers. All stakeholders should be involved in writing agreements and developing holistic goals.

AGRARIANTRUST.ORG/PRINCIPLES

"Civilization... or that which is so-called, has operated two ways: to make one part of society more affluent, and the other more wretched."

"It is a position not to be controverted that the earth...was, and ever would have continued to be, the common property of the human race."

"The plan I have to propose...is to create a national fund, out of which there shall be paid to every person, when arrived at the age of twenty-one years, the sum of fifteen pounds sterling, as a compensation in part, for the loss of his or her natural inheritance, by the introduction of the system of landed property."

—Thomas Paine, From *Agrarian Justice* (1797)

Three Millennia of Agrarianism

Rose Karabush

Agrarianism is either a philosophy that assigns farmers a privileged place in society, or a movement that attempts to redress issues such as land distribution or the economic status of farmers. Agrarian ideals have fostered hope, dignity, and justice, but they have also contributed to the devaluation of "non-agrarian" land management, the dispossession of indigenous peoples, and some very outdated agricultural policies. To build a more just agrarian future, we must recognize the power of agrarianism for destruction as well as for regeneration.

The School of Tillers, China, 770–221 BCE

A wise ruler tills the land together with his people to make his living."
—Xu Xing

The School of Tillers flourished during the "Hundred Schools of Thought" period in classical China. As opposed to their contemporary rivals (such as the Confucians and Legalists), the School of Tillers advocated egalitarian self-sufficiency rather than a division of labor, and advised kings to lead through instruction rather than through punishment. The School of Tillers rejected war-making and capital punishment, favored fixed prices on all goods rather than market competition, and believed that all people had a natural propensity to work the land.

Sabbath Years and the Jubilee, Judea, 600 BCE–135 CE

"And the land shall not be sold in perpetuity; for the land is mine..."
—Leviticus 25:23

Among the many laws for the Hebrew people in the Bible's Book of Leviticus, quite a few concern agriculture. These instructions mandate a Sabbath Year once in every seven, when agricultural land must be left fallow, and also decree that the year after every seven Sabbath Years—the 50th year—would be the Jubilee Year.

Particularly in the ancient world, calamities such as war or bad weather often resulted in peasants incurring debt, and many eventually lost their land and even their freedom. On the Jubilee year, all peasants who had lost their freedom due to debts were emancipated, and all land would revert to its original owners. In this way, the failures or misfortunes of one generation wouldn't be passed down in perpetuity.

Lex Sempronia Agraria, Rome, 133 BCE

"...they fight and die to support others in wealth and luxury, and though they are styled masters of the world, they have not a single clod of earth that is their own."
—Plutarch

The Lex Sempronia Agraria was a land reform bill passed by the Roman politician Tiberius Gracchus. Gracchus aimed to return land to the peasant farmers who had lost theirs (in situations similar to those addressed by the Jubilee), and had been forced to move into the cities. Land would be confiscated from those who owned more than the equivalent of 311 acres, and returned to poor families in parcels of about 19 acres.

This redistribution of land was partially intended to improve the food security of Rome, since the large landowners of the day exploited the work of tenant farmers and slaves to grow cash crops such as wine and oil rather than the staple crops that had been grown by smallholding peasants. However, the motivation for this bill went beyond the obvious appeal to its beneficiaries or the need to correct a grain shortage: by creating more landowners, the Lex Sempronia Agraria also generated more taxes and more military conscripts for Rome.

Agrarian Theories of Wealth and Property, Colonialist Europe, Late 17th–18th Century

"…whatsoever he tilled and reaped, laid up and made use of, before it spoiled, that was his peculiar right;"
—John Locke

Agrarian theories of wealth and property were integral to the birth of modern economics and theories of natural rights. The French Physiocrats of the 18th century believed that the wealth of a nation came solely from agriculture and glorified those who worked the land. However, they were not necessarily concerned with the distribution of this wealth, or the well-being of the individual farmer. They were the first to propose a "laissez-faire" policy, and paved the way for the classical school of economics (Smith, Ricardo, Mill, etc.).

John Locke, who predated the Physiocrats, advanced a very agrarian theory of property rights in part to justify English colonialism in America: that the ownership of property comes from the application of labor to nature. At least in territory which remained in the "state of nature" (i.e. pre-colonial America), "improving" land conferred ownership of it. What Locke meant by "improvement" was traditional European agriculture: tilling, reaping, etc. He was careful to specify that Native American hunting or gathering entitled them only to the meat or fruit that they harvested, not the land itself.

Jeffersonian Democracy and The Agrarian Myth, United States, 1790s–1820s

"Cultivators of the earth are the most valuable citizens. They are the most vigorous, the most independent, the most virtuous…"
—Thomas Jefferson

Jeffersonian democracy refers to the political outlook of the Democratic-Republican party, which favored the interests of the idealized, self-sufficient yeoman farmer and promoted equality of political opportunity. They pushed for westward expansion to obtain more land for these "valuable citizens."

Jefferson's vision of the virtuous farmer as the foundation of American democracy has had a political life far longer than the actual prevalence of self-sufficient yeomen. This vision has attained the status of myth in the culture of the United States. It is not necessarily untrue, but its importance goes beyond mere fact (contributing perhaps, to continuing illusions about the largest beneficiaries of U.S. agricultural policies).

Distributism, Europe, Late 19th–Early 20th Century

"Three Acres and a Cow."
—Distributist Slogan

Distributism is a an economic ideology which comes from the doctrines of Catholic social teaching, especially those of Pope Leo XIII and Pope Pius XI. Proposed as a middle way between socialism and capitalism, distributism posits that widespread distribution of the means of production is key to a just society. Distributists believe that private ownership of land is a god-given right, and that it creates an ideal situation where people can make a living without relying on the property of others.

Agricultural Collectivization under Stalin and Mao, USSR and PRC, 1929–1961

"The changeover from individual to socialist, collective ownership in agriculture… is bound to bring about a tremendous liberation of the productive forces."
—Mao Zedong

Unlike most of the above examples of agrarian movements, the agricultural reforms attempted during Stalin's first five-year plan and Mao's Great Leap Forward did not promote the independent, self-sufficient farmer, nor were they seeking wider distribution of land ownership. These were, however, both efforts to redress social and economic ills through land reforms and agricultural "modernization." In both instances, collectivization—through mismanagement and unforeseen consequences—led to catastrophe. Millions died by famine in both countries.

Farmer Direct Action, Worldwide, Middle Ages–Modern Day

While most types of agrarianism addressed here are top-down or intellectual, there have been innumerable grass-roots movements. The desire of peasants, serfs, and farmers for a better agricultural system has motivated revolts and protests from the Middle Ages to the Industrial Revolution and beyond.

Direct action to improve the lives of farmers and to address inequitous land distribution continues today all around the world, and many of these organizations seek to reform conceptions of appropriate agrarian land management, improve their governments' agricultural policies, and support justice and well-being for indigenous peoples.[1]

1   Find Agrarian organizations, and more, working near you on the Greenhorns Resource List, at: http://www.thegreenhorns.net/category/resources/resources_list

EDITOR'S NOTE In 1978, the American Agriculture Movement marched on Washington DC with a tractorcade to protest foreclosures and demand parity to match market prices to the costs of farming.

"**Wherever we turn, to Asia, Europe, or Africa, we shall find the same story repeated with an almost mechanical regularity. The net productiveness of the land has been decreased. Fertility has been consumed and soil destroyed at a rate far in excess of the capacity of either man or nature to replace. The glorious achievements of civilization have been built on borrowed capital to a scale undreamed by the most extravagant of monarchs. And unlike the bonds which statesman so blithely issue to—and against their own people—an obligation has piled up which cannot be repudiated by the stroke of any man's pen.**"

—Paul Sears, from *Deserts on the March*

	International Morse Code	Semaphore	Manual Alphabet			International Morse Code	Semaphore	Manual Alphabet
A	·—			N	—·		
B	—···			O	———		
C	—·—·			P	·——·		
D	—··			Q	——·—		
E	·			R	·—·		
F	··—·			S	···		
G	——·			T	—		
H	····			U	··—		
I	··			V	···—		
J	·———			W	·——		
K	—·—			X	—··—		
L	·—··			Y	—·——		
M	——			Z	——··		

April

Quality of Life

Chapter 4

Field Notes From
A First Season Apprentice

Farming five acres of mixed vegetables
in the Hudson Valley, New York, 2013.

Audrey Berman

April & May

4.6.13
Cluster flies wake up in Spring and feed on earthworms.

4.8.13
To see if soil is dry enough to work, pick up a handful, squeeze into a ball and drop from your waist. If it breaks apart that means it's okay to work.

4.9.13
Ear protection should block out sound at a minimum of 29dB.

4.20.13
"Field factor"—drop a little extra seed when direct seeding in case there is poor germination.

4.22.13
It's ideal to cultivate at the "white thread" stage, before weed roots have had a chance to really take hold.

4.24.13
A firm seed bed is great for crop germination but also good for weeds. Cultivate pathways to prevent weeds.

4.29.13
Laying down a hay mulch on tomatoes and eggplant will increase tilth for the following year, even though weed seed will be imported. It is okay because hay seed is not a long term weed. Once it's cultivated enough in a season it will be gone.

5.6.13
Gold Star Chickens - their small size means they eat less to make the same amount of eggs as a full size chicken, people friendly, great egg layers in both cold and hot temperatures.

May & June

5.7.13
When transplanting summer squash, be careful not to disturb their roots, as they are especially sensitive.

5.13.13
Don't grow what your neighbor is growing, the market is saturated with easy to grow vegetables.

Randall Lineback Cattle - triple purpose breed; milk, meat and draft.

5.14.13
Weed onions well because their leaves will never shade out weeds and they stay in the ground a long time. It's best to stay on top of them.

5.22.13
Best not to touch solanaceous and cucurbits after rain, it increases your risk of spreading disease.

Always harvest the newer beds first.

5.30.13
"Dust Mulch" - after rain, to preserve moisture in the soils it is best to cultivate so as to create a layer of dust mulch. This will prevent the capillary action and evaporation of water on the soils' surface. Do not try this right after rain but after the soil has dried out a bit.

6.3.13
Sheep netting works well as fencing for chickens.

6.21.13
When harvesting parsley, don't remove more than a third of the plant, otherwise it will bolt from the stress.

Note to future self: use metal sprinklers for irrigation because the PVC ones break a lot! In the hot summer months, germinate lettuce in a shady place.

June, July & August

6.25.13
Harvest basil very close to distribution, it turns brown quickly.

7.3.13
When using row cover, shoot for 20' minimum width to prevent flea beetles from eating brassicas. They will get under cover no matter what and eat what is closest to the edge. If you only have one row of cover then they will eat the entire row, meaning that you should try to plant a few rows of brassicas together.

When finger weeding in not ideal conditions, move weeds outside the spaces between plants so that if they re-root, you can always kill them next time you cultivate.

7.23.13
Dry soil is also not good to work up with tractors, you can destroy the soil structure.

Kranzle pressure washer is the best.

Buckhorn bins are great for storing vegetables in the cooler as well as for harvesting in the field.

Pick summer squash, zucchini and cucumbers every two days in the summertime.

It's okay to leave non-brassicas outside of the greenhouse at night. Close up greenhouse at 8:30 PM at the latest to prevent pests from getting to the transplants.

8.9.13
Debris leftover from a cover crop is good for preventing a crusty layer on the topsoil from forming.

Ode To Trade

Schirin Rachel Oeding

People say that we owe each other everything.
I'm not sure what to think about that.
Walking home along the dirt road last night,
I didn't feel indebted to anyone.
Maybe debt is the wrong word.
Maybe we should be thinking in terms
of something more neighborly,
something that doesn't cost us
an arm and a leg.
What would I like to give you today?
I would like to give you a good long look
at the yellow crocuses coming up
along the house wall.
And the smell of mud,
and the sunshine that's yours for the taking.
Next time we meet,
and you write me a check,
I'll use it to kindle the fire under my
sugaring pan.
I'll send you a jar, when everything is said and done.

On my way home,
I had almost everything I need
—we'll trade, let's say,
and stay on even ground that way.

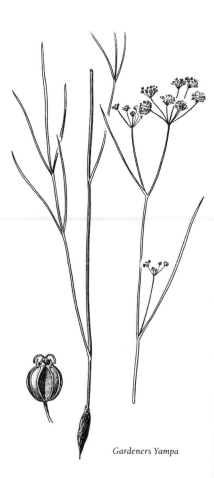

Gardeners Yampa

Health Insurance for the New Farmer

Audrey Berman & Rose Karabush

Sustainable farming is itself an investment in health and holistic wellness. The health benefits we get from farming are many, but farming also presents hazards not known by most workers. In order to protect our health and the health of our families, most of us will need to purchase medical or wellness services at some point in our lives. Health insurance is the most common way to ensure that we can afford the often exorbitant prices of medical care in the United States. This article is a bare-bones overview of health insurance: we're not public

health experts, and the durability of this advice is far from certain. But in the spirit of farmers everywhere, we're gonna do what we can with what we've got.

First, some basic vocabulary.

When comparing health insurance plans, there's a couple of specialized terms that refer to who pays what, and when. A premium is a fee that you pay for your insurance coverage, usually monthly, no matter what. There are co-pays, which are fixed prices that you pay for routine

services such as yearly checkups or prescription drugs. Then there are deductibles and coinsurance, which apply mostly to non-routine care: if you get surgery, blood tests, or hospitalization, you will foot the bill until you reach your deductible, which is a dollar limit usually calculated annually. After this point, the insurance will start paying at least part of the bill, but you may still be required to pay coinsurance, which will be a percentage of the total bill after your deductible has been reached. Some plans may have a total out-of-pocket limit as well,

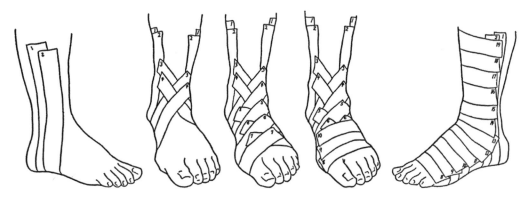

which sets a yearly maximum on what you pay for covered health services. This may or may not include co-pays and coinsurance, and never includes your monthly premiums. Terms such as HMO and PPO describe how your insurance plan limits the service providers that you see—in an HMO, you choose a primary care physician, who must refer you to other services, and in a PPO you can choose within a network of approved providers.

If you are currently employed, serious on-the-job injuries will often be covered by workers' compensation insurance. The laws governing this change state-to-state, so it's best to do your own research into what's required of your employer or to figure out what your own policy covers.

You probably have heard of the Affordable Care Act (ACA)—also known as "Obamacare"—and the penalty for the uninsured that started last year. There are a few exceptions to this penalty, and one is if you make less than a certain amount of money per year ($10,000 for an individual, $20,000 for a family).[1] Of course, if you make less than that, you probably also qualify for free health insurance—i.e. Medicaid—so either way, read on!

So, what are the health insurance options for a new farmer today?

They vary widely based on your situation. The following advice is intended mainly for the self-employed or those who don't have access to health care through their employer or a parent's employer (for those under 26). In most cases, employer-sponsored health insurance will be the best deal, because of the health care market structure and tax incentives. Starting in 2015, some small businesses will be penalized if they fail to provide basic health insurance to their employees, but this does not apply to seasonal or part-time workers.

No matter what your employment situation, Medicaid is available to those who fall under a certain annual income threshold. This varies state to state, but in 2014 the ACA lowered the threshold for Medicaid in many states (mostly blue states) to at least $15,856.20/year for a single person, or $32,499.00 for a family of four.[2]

Even if your income is above this threshold, you may qualify for the subsidized insurance plans which are available to people making between 100% and 400% of the

Federal Poverty Level. You can determine your eligibility for both types of plans as well as compare private non-employer-sponsored plans on your state's health insurance exchange website. If your income falls within the eligibility range for both Medicaid and the subsidized insurance plans, you should be able to choose between them. In general, Medicaid is free, but is not accepted by many physicians or hospitals. The subsidized insurance plans require some monthly premiums but may provide better or more convenient coverage. If you are a seasonal worker, applying for Medicaid in particular may be tricky during the on-season, since they calculate your annual income using your current monthly income. You should be able to straighten this out (perhaps by contacting a Navigator, whose job it is to help people sort out their health insurance under the ACA), but it might be easier to simply apply during the off-season.

If you have a choice between different health insurance plans, there are many factors to consider. Outside of Medicaid and subsidized plans, the cheapest insurance in terms of premiums usually has a very high deductible and limited coverage. All health insurance sold on the new state Marketplaces will cover a basic set of benefits (Essential Health Benefits) as defined by the government, and they will also be rated by actuarial value, which is the total percentage of health care costs that the plan will pay for the average person insured. Especially if you live in a rural area,

1   The Henry J. Kaiser Family Foundation, "The Requirement to Buy Coverage Under the Affordable Care Act," http://www.kff.org

2   Calculated based on 2013 Federal Poverty Levels for the 48 contiguous states. American Public Health Association, "Medicaid Expansion," http://www.apha.org/ CMCS/CAHPG/DEEO, "2013 Poverty Guidelines," http://www.medicaid.gov/

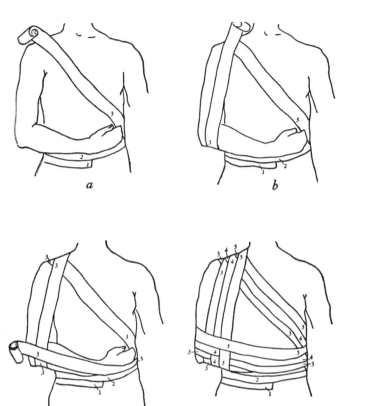

a

b

c

d

and there are even farmer-specific co-ops in states like Wisconsin[3] and California.[4]

Now, thanks to loans provided under the ACA, co-ops are becoming more widespread, and readers of the *New Farmer's Almanac* may appreciate the cooperative spirit and lack of profiteering from these new insurance providers. It remains unclear, though, how these new co-op plans will stack up against the subsidized plans on the state exchanges in terms of price and value.

What can a forward-thinking farmer do?

The U.S. health insurance market has changed dramatically just in the past few years. One option to improve the health insurance possibilities that will be available to the new farmers of the future is to engage in what is sure to be a long-lived discussion on federal and state policy. Or, if you like the idea of the health insurance co-ops but your state doesn't have one yet, help start one up![5]

We would also like to propose a few non-governmental possibilities for farmers and lay-people. Though not

another important factor will be the provider network—whether there are covered services in your area. Most of your evaluation of a plan's covered benefits depends on what you expect you might need, so be aware that many do not cover dental or eye care, and they may vary in their coverage of preventative services, mental health services, alternative health care, childbirth, and children's medical services. Coverage of specialty care (such as OB/GYNs, for example) may also vary, so if you have a particular physician with whom you'd like to stay, be sure to check whether

they are in your network and will be covered. When considering your options for health care, no matter what your insurance coverage, you may also want to check out any local Federally Qualified Health Clinics or Planned Parenthood clinics, both of which provide some free or sliding-scale health services for men and women.

There is another possibility for affordable non-employer-sponsored health care—health insurance co-ops. In various incarnations, health insurance co-ops have been around for years,

3 "Farmers' Health Cooperative of Wisconsin," www.farmershealthcooperative.com
4 See: "United Agricultural Benefit Trust," www.uabt.org
5 Michael Meulemans, "How to Launch a Health CO-OP Plan," www.insurance.about.com

a replacement for health insurance, one inspiring way to deliver health care to under-insured populations is that demonstrated by the O+ Festival, which takes place yearly in Kingston, NY and San Francisco, CA. During this three-day festival, artists of all sorts exchange their work for free healthcare from volunteering medical providers. We envision a farmer version, in which a community's foodshed might get together and hold a yearly harvest feast and festival, and farmers and local food systems workers would donate skills or surplus in exchange for medical services.

Another possibility is to extend the lessons of our much-beloved CSA model into medicine and wellness: have community members commit to a certain level of health and wellness services in order to receive discounts and support their providers (take for example the Brattleboro Holistic Health Center of Vermont, which is a co-op that offers discounted monthly plans for alternative healthcare). This may seem like a very minor change, but we see it as one step on the path to an essential shift in perspective—a repersonalization of our relationship to health care providers.

Like many aspects of our lives, we think health care could be improved by a focus on local communities and human connection. This could mean community financial support for those who have spent the time and money to acquire the skills to care for us, it could mean embracing traditional and non-disciplinary peer-to-peer health support like that of doulas or herbalists, or simply organizing community events such as a fitness challenge or a massage circle. Perhaps we might even imagine a time where communities could support and build relationships with their own traveling doctor, or promote wellness through programs in thriving new Granges. As farmers, we help steward the most fundamental levels of health in our communities: clean air, clean water, nourishing food. In most cases, we've already invested deeply in the strength and well-being of our communities—working towards more sustainable and rewarding health care systems is very much our business. So let's take care of ourselves and our families, the land and each other, and keep trying out better ways of doing so.

HOUSEHOLD HINTS.

THE USES OF THE LEMON. — As a writer in the *London Lancet* remarks, few people know the value of lemon juice. A piece of lemon bound upon a corn will cure it in a few days ; it should be renewed night and morning. A free use of lemon juice and sugar will always relieve a cough. Most people feel poorly in the spring, but if they would eat a lemon before breakfast every day for a week, — with or without sugar, as they like, — they would find it better than any medicine. Lemon juice used according to this recipe will sometimes cure consumption : Put a dozen lemons into cold water and slowly bring to a boil ; boil slowly until the lemons are soft, but not too soft, then squeeze until all the juice is extracted, add sugar to your taste, and drink. In this way use one dozen lemons a day. If they cause pain, or loosen the bowels too much, lessen the quantity and use only five or six a day until you are better, and then begin again with a dozen a day. After using five or six dozen, the patient will begin to gain flesh and enjoy food.

Javelle Water is often recommended for stains on white goods. Use carefully as it rots the material. Oxalic acid solution should be used after it and the material on which it is used should be rinsed thoroughly. It must not be used on silk or wool.

The United States department of agriculture gives this method of making Javelle Water: Dissolve one pound of washing soda in a quart of cold water. To this add a quarter of a pound of bleaching powder (calcium Hypochlorite). Filter this liquid through muslin and keep tightly sealed when not in use. In using it, stretch stained portion over bowl and apply liquid only to spot. Immediately apply Oxalic acid solution and rinse well. Several applications may be necessary.

A Deer

Schirin Rachel Oeding

This morning
walking with the dogs in the woods,
I saw some drops of blood
which had saturated the snow,
bright against whiteness.
A little way down the path,
under old white pines and hemlocks,
in the ditch
was a dead deer.
Its ribcage was exposed and
lay in a reddened circle like a
giant insect,
striped red and white
with bone and flesh.

Later someone told me it had been
hit by a snowmachine,
but this morning mechanics seemed
far away and I felt
closer to the deer.
In a day there will be tracks everywhere
and just tufts of hair
and bone remaining.

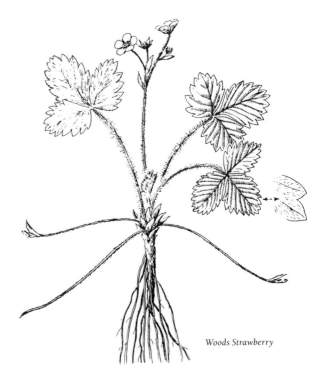

Woods Strawberry

Before We Turn To Dust:
New Expectations As A Farmer Goes To Seed

Khristopher Flack

It was around the 13th consecutive day of work early in my third season of vegetable farming when I considered how long it'd been since I'd seen a movie, called a friend, gotten past the first lines of a poem, read more than ten pages of a book, or just sat and did nothing. I rolled my shoulders back in line with my ears and stood up

straight. I was one of four people in charge of the farm, and I didn't feel like I had much control.

I'm sure all of us have been told by a more experienced farmer that "this is the way it is." But the contradictions between the ethos of how we love our land and the

Willa Mamet *Dandelions*

encouraging a new generation of growers to raise food at the scale of their needs accordingly, it's important that the potential of community, school, and other small-scale gardens not be misrepresented as a novelty. These spaces are not separate from the "real" production power of mightier farms, but in fact they represent a critical mass that could decentralize the farmers burden and make food production more sustainable for all. Rather than suggesting everyone become a full time farmer, municipal-scale agriculture questions the wisdom of farming as a full time profession at all. We don't need farmers who raise commodities; we need human beings who raise food.

Cultivating this new farmer also means razing our nationalized image of The Farmer. Perhaps no other character in Americana carries such a deeply romantic heroism. We expect him (and it is always a "him") to be up at every dawn; we expect him to toil through every dusk; we expect him to be tight-lipped and sullen; and we expect him to face incredible adversity regularly. We also expect him to pull through and manage to feed us all.

practice of how we love ourselves have twisted our bodies into question marks. They beg us to question the way it is, and look for a way forward that irons out those wrinkles in our definition of sustainability.

We can begin making more space for ourselves by making more space in the job description. If 100 hours/week and endless self-sacrifice aren't enough to get the job done, then there's not something wrong with the worker, there's something wrong with the job. The nature of this dilemma is even clearer when we consider the farmer's burden in the statistical context of national food production. 1% of 314 million people is expected to raise quality food for the other 99%.

This kind of imbalance in the economy was enough to occupy streets around the world. We can alleviate the same imbalance in our food system by occupying lawns, lots, rooftops, and any other arable space in our communities with gardens. In doing so we shatter the most constrictive shackles on the farmer—centralization and scale—while

But this character is a farmer, not a person, and while we should be proud of our tenacity, we can never expect ourselves to be healthy as long as we continue to buy back our own anxiety, depression, and stress when it's sold to us as valor. If that makes us less American, then let that sharp idea cut another fiber on the rotting rope of patriotism. These national-scale institutions deserve to be left in the past they came from.

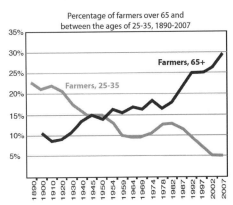

Percentage of farmers over 65 and between the ages of 25-35, 1890-2007

As the food movement matures through its adolescence, this would be the real coup: a generation of growers, working on a smaller scale, supported by lots of micro scale growers nearby, all of whom take pride in maintaining the well-being of their literal and metaphorical soil. Particularly now, when the initial righteous high of being a greenhorn may be plateauing into the demanding reality of what we've taken on, it's important for young growers to consider our responses to those demands proactively. We must cultivate not only a movement of additional hands, but a movement of whole, balanced, healthy people whose hands will be able to continue working the earth for more than their first few years of excitement.

If we don't, we won't have a new generation of growers; we'll have a generation of suns who cook themselves before they rise, and move on to less damaging lifestyles. Granted, there are people who genuinely enjoy the rough shake of farming as it is. But certainly even those folks wouldn't mind something a little more fair. And fairness starts with giving ourselves the freedom to cultivate lives as diversified as our gardens.

If we're serious about defining a new generation of sustainable agriculture, we must consider the farmer's personal well-being to be part of our ecology, and be willing to make these personal shifts along with systemic shifts in expectations and scale like those outlined above. Until we are eating food that did not require a farmer to till their soul to dust, our food supply is just as poisoned as the most genetically modified, chemically-dependent system we can imagine.

The Husking Bee

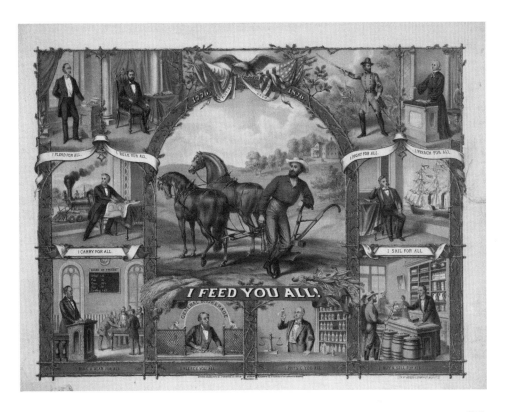

REFLECTION

I think I shall decide to stay
   Here in a field with a fence around,
Sowing some oats and making some
   hay
   And learning the ways of a piece of
   ground.

There will be time to watch the birds
   Perch on the sky, a wavering shelf,
While I am thinking important words
   To say to men who are like myself.

I shall have this to recall when green
   Seasons are gray and days are
   thin:
The infinite wonders that I have seen,
   And the curious person I have been
      —James Hurst, Iowa farm boy.

Difference Between Farm Bureau & farm bureau Explained

Dear Kansas Editor:

This page with the three-column cut of clippings taken from Kansas newspapers has been prepared to point out to the newspapermen of the state how without realizing it they are contributing to the perpetuation of the confusion and misconception in the popular mind of who and what the Farm Bureau, the farm bureaus, the extension service and county agents are.

Let us first make no bones about who the Farmers Union is and what our interest is in desiring a clarification among the citizenry of Kansas, particularly farmers, of the differences between the Farm Bureau and the Extension Service of the Department of Agriculture.

The Kansas Farmers Union is officially the Kansas Division of the Farmers Educational and Cooperative Union of America. We are a general farm organization of family-type farmers.

The American Farm Bureau Federation is likewise a general farm organization. It's present president, Edward C. Neil, has stated that his organization represents large commercial farmers.

Naturally, the interests of two such organizations as the Farmers Union and the Farm Bureau are often divurgent.

We are not seeking the support of the press for the Farmers Union's program as it stands in opposition to that of the Farm Bureau's.

You will note that the underlined words in the clippings are "Farm Bureau" with each word beginning with a capital letter. This is imporant because:

The extension service is an educational agency, supported by all taxpayers. State laws of Kansas specify that whenever a county "farm bureau" (and note that here the words do not begin with capital because bureau is used in the sense of a department or office or association) composed either of a certain number or else a percentage of the farmers of a county that the county and the Kansas state college shall jointly support a county agent to serve all the farmers of that county.

Now we have a Farm Bureau which is a general farm organization and we have farm bureaus which are simply organized to meet the requirements of the law and bring county agents and home demonstrations agents into a county for the purpose of giving instructon in agriculture and home economics to the people of said county through practical demonstrations and otherwise.

Everytime in the news stories pictured on this page that the "Farm Bureau" is written, the farm bureau or extension service is meant.

Is it not good journalism to keep the record straight? It is as much a mistake to call the farm bureau the Farm Bureau as it would be to call a believer in democracy a Democrat. While they may be both they are not one and the same, and no politics meant.

A means of further clarifying matters would be to refer to the county agent, his office and activities as extension service, using "Farm Bureau" only when referring to the American Farm Bureau Federation.

The newspapers of Kansas can perform a real service in making this honest clarification of terms. The American Farm Bureau Federation will lose nothing, the extension service will gain by being responsible for only its own actions and will get the credit for services which it normally conducts but which is erroneously credited to "Farm Bureau" by confusion of the terms in the press.

You will not be building the Farmers Union or in any way supporting it by using the terms we have suggested. You will only be practicing more accurate reporting.

Sincerely, The Editors, Kansas Union Farmer

Honey Bees

Schirin Rachel Oeding

Talk is cheap,
don't you know?
Sentences unintelligible
and the small movements of your mouth
impossible to discern by any but the
most skilled reader of lips.
While we sat around for a few hours
growing uncomfortably square in our
square-bottomed chairs,
I watched the sun play across the curve
of your cheek, and the small scar above your
left ear,
and I learned by heart the turn of your eyes
when you talked about togetherness,
and I memorized the rhythm that your
nervous fingers tap-tapped onto your thigh.
I have stopped hearing you,
I am sorry to say.
It was not my choice, you should know—
my ears have become attuned to
other melodies.
Singularity no longer suits them—
they love the sound of a hundred bees
in a field of pink poppies.

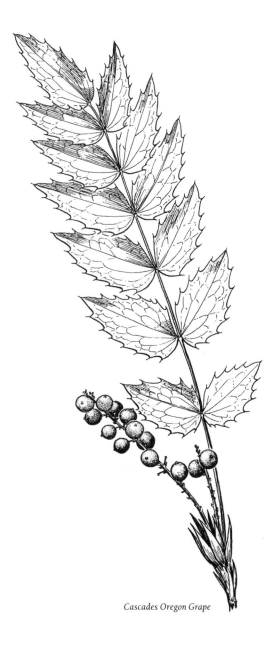

Cascades Oregon Grape

Backwash: A Study On Dirt, Crass, And Fermentation

Natsuko Uchino

Backwash refers to any backflow of air or water. Originally, this referred to the receding current after a wave had hit the shore. In American slang, the word is used to designate the liquid remaining at the bottom of a beverage—a mixture of the liquid and saliva—and is considered a disgusting thing. In Argentina however, people share the same straw all the time. In fact when sharing yerba mate it is a rule that you always use the same gourd and straw—and thus drink each other's backwash.

Many of my fellow Japanese wear masks in streets and in public transportation as protection against pathogens and dust. And yet these same people share their bath water, letting public bath-houses go at least a day before changing it.

So what's gross and what isn't?

It's interesting to note that what's gross does not always match what is not clean. One is a socially-determined parameter, the other is a scientific description of the sterilized. And in this battle against pathogens fueled by fears of contamination, there has been a fervent enthusiasm for disinfecting anything from a minor cut to the kitchen counter.

A sterile environment is perhaps clean, but is it resilient? It strikes me that the same word, sterile, is used to refer to one that is unable to procreate. And we also use the same word, culture, to designate bacteria or arts. A culture that is sterile is stagnant. A live culture is generative. With fermented foods, if the culture is properly fed, it will grow endlessly.

A young baker told me about the history of yeast. In the days of the communal wood-fired oven shared by the village, there was also a collective batch of live yeast culture. People would take turns, baking their bread, keeping up with the fire, handling and feeding the yeast. In the process, neighbors would introduce their own bacteria into the culture and thus the people developed a stronger, more diverse immune system.

This is similar to Sandor Katz's notion of "eco-immuno-nutrition" which recognizes "that an organism's immune functions occur in the context of an ecology, an ecosystem of different microbial cultures, and that it is possible to build and develop that cultural ecology in oneself through food." The system is a rich microbial world that will be annihilated by the antibiotics and the bleach. If this happens the system is left bare, deplete, defenseless.

Louis Malle also complicates this in his series Phantom India. He describes how some Indians scrub themselves "relentlessly" in the temple pool, leading to the possible conclusion that they are some of the "cleanest people in the world." The water however, is filthy, and the scrubbing is actually an "element of worship in a... culture founded on criteria of pure and impure."

We see then, that cleansing is also based in belief.

We live in waste, but there are many delusions we subscribe to in order to support this manufactured society. The collective urban delusion is that water flows from the tap, trash disappears from the curb, the toilet flushes out of sight, light turns on from the switch, and heat comes from the pipe. The how is removed from me. We are left with a blank in the process of consumption.

In Paris there was a memorable couple of days in the late nineties when both the workers that refurbish the ATM machines and those that pick up the trash went on strike at the same time. During those few days, all of Paris was amidst rotting garbage and nobody had any cash. Maybe that was the best prediction about what the future is actually going to look like.[1]

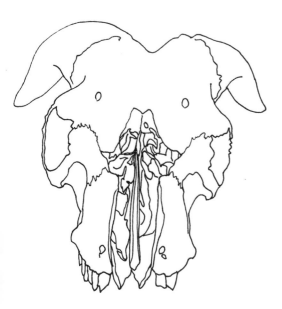

PRAYER

Lord, I shall be very busy this
Day and may forget about Thee;
But please dear God, don't forget
About me.
—Sir Jacob Astley

"BUT. . ."

I am the "butter." I get my name by injecting the word "but" into what otherwise would be high compliments for people and things. "Yes, he's a fine young man; one of my best friends. I like him a lot, but . . ." "Sure, he's a fine preacher, and the church likes him, but . . ."

"Yes indeed! America is certainly all right, but . . ." I'm a crack shot, I'm not going to stand by and let anybody or anything get away with unlimited approval! There are no closed seasons on compliments. I shoot them down anytime with my distinguished slander-gun. I never miss a shot!
—Western Recorder

GOOD TEETH FOR HEALTH AND LOOKS

Your smile is your fortune. The beauty and effect of your smile depend on your teeth. If your teeth are yellow and decayed, your smile will have no sparkle. Without teeth your face becomes sunken and lined.

For clear, well-pronounced, pleasant-sounding speech you need your teeth . . . you need them in good shape and evenly spaced.

To enjoy your food and to chew and digest your food properly, you need your teeth . . . in good working order.

A grown-up's teeth usually reflect the care that was given to the first or "baby teeth." Good care of children's teeth helps to insure sound teeth in adult years.

From baby's very first tooth on, good dental care is important because the baby teeth
. . . help shape the jaw for the later permanent teeth.
. . . are necessary for chewing food.
. . . are needed for proper development of the jaw to insure attractive shaping of the face.

If baby teeth are neglected, allowed to decay or are lost too early, the permanent teeth may come in crooked and badly spaced.

The keystones to the dental arch are the four permanent teeth (called molars) that come in at six years of age. These teeth act as guides to all the later permanent teeth. Other teeth may grow in crooked if the

"keystone" molars are lost. So take good care of them.

Decay is the greatest enemy of your teeth. One of the chief causes of tooth decay is the way sugar acts after it reaches your mouth. Action begins when you suck on a piece of candy or eat a sweet dessert. The sugar, or sweetened foods you eat, makes it possible for the many germs already in your mouth to start working. They ferment the sugar and produce an acid. It is the acid that eats into the enamel and starts the decay. Decay will destroy the teeth and often bring on the pain of a toothache.

You can have good teeth by beginning dental care early. The child's first visit to the dentist should be at 2½ or 3 years and everyone should see his dentist regularly throughout life. Often the dentist can fill small cavities before they get big and start to ache.

To help build sound teeth, children must have the proper foods: milk, cheese, meat, leafy green and yellow-vegetables, whole-grain cereals, eggs and fish liver oils. To keep your teeth in good condition, eat less sugar. Cut down on candies, sugary desserts and soft drinks.

Brush your teeth immediately after meals to stimulate the gums and to clean the teeth. For tips on the right way to brush your teeth ask your dentist or dental hygienist.
—N. Y. State Dept. of Health

---

1 Backwash was originally written in 2009 for a zine of the same title. It has been amended and edited for the New Farmer's Almanac.

Notes on Drawing

Aimee Good

I

My kind of drawing begins with the body—proximity to people, place and landscape.

II

My relationship to drawing began as a child sitting on the fender of a green Oliver tractor next to my father, Tom, a potato farmer in Aroostook County, Northern Maine. He taught me to see—far and wide— the arc of the sky as it bent to meet the deeply jagged, textured tree lines of Maple, Fir, Pine, Hornbeam, Birch, Poplar, Elm. I watched drawing unfold in the linear grooves cut in the freshly turned earth as he planted grain, measuring raised relief and line weight in the super straight hillocked rows of potatoes that he made sure were planted perpendicular to the horizon line of the field. My understanding of perspective came when he would pick a tree as a focal point on the field edge, perhaps a bright white birch, to anchor his navigation over the constant rolling surface of the earth—repetition and pattern.

(pause)

If a line is a collection of moving points, then we were a line. The body as a fixed and moving point upon the landscape. A member of a family, a community, a crew of people working the farm. Our bodies as we worked picking potatoes, picking rocks, hoeing or weeding became moving points, an animated line constantly shifting under the cloudy sky on acres of textured surface of topsoil.

In the house, my mother Rachel drew with her sewing machine. Her hands took the flat painted fabric and transformed it into my favorite clothing year after year.

Fitting my body, lovingly shaping, tucking, marking, tracing. Miles and miles of threaded line passing through her fingers.

III

I think about the traces of my tires on the long line of road between New York, where I live and work;

Boston, where my creative life was established; and Maine where I farm 30 acres organically in collaboration with my family. The etched groove, a 570 mile long line—a drawing. I trace a mark, circle the spot, points of contact with others, using the vocabulary of drawing to actively think through my life—my relationship to self and community. I think about how to activate the farming community I grew up in, to be of service in forming a healthy, diverse agricultural future in Northern Maine and New York— how to create cultural engagement of mark making, to make a mark on and in this place.

IV

My early life on the farm also exposed me to the language of collaboration, the art of working together to achieve something larger than one man or woman could pull off alone. Led by my tireless father, his oft repeated phrase, "Get er done!" was and continues to be the Good Farm rally cry for motivating.

Through collaboration we have an opportunity to learn a different facet of what it means to be generous.

Trust, intimacy, adaptation, communication, experimentation are the major tools for collaboration. These tools reshape and amplify the creativity and catalyst for change that already exists within each of us and within our communities.

I advocate for collaboration—not always, but at specific junctures in our lives when the agitation of opposing visions is needed to sharpen perspective. Different viewpoints clarify the larger intention, and in being asked to collaborate we are asked to bring not just agile minds and skilled hands to the engagement, but supple hearts as well.

I am interested in supple hearts.

(pause)

Together we envision future and form generative unlikely alliances.

We construct social bridges. We become more fearless.

V

Wendell Berry, the farmer philosopher, writes about familiarity of place. Farming is about the familiar—returning over and over to places of infinite discovery.

These are his words:

"Living and working in the place day by day, one is continuously revising one's knowledge of it, continuously being surprised by it and in error about it. And even if the place stayed the same, one would be getting older and growing in memory and experience, and would need for that reason alone to work from revision to revision. One knows one's place, that is to say, only within limits, and the limits are in one's mind, not in the place."

In our rapidly changing world, the intimate familiarity of place is an embrace that remains constant—magical.

Drawing brings us full circle.

(end)

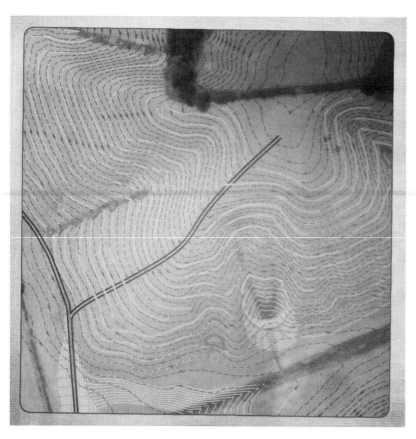

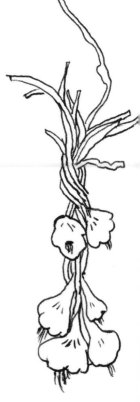

Sap of maple & birch, drunk by the gallon

Pancakes fried thick & buttered, stacked high, with syrup from the trees

Pancakes fried thin & wrapped round every good thing

The wet, orange yolk of soft boiled eggs drizzled hot over buttery bread

Mugs of thin goat yogurt to whet our lips

Hard sausage with cheese on brown bread

Smoked fish, entire, with the eyes

Salt fish with dill, with cream sauce, with mustard

Broth over meat dumplings with sour cream

Cucumbers in brine with oak leaf, garlic, dill

Tomatoes in brine with currant leaf, garlic, dill, honey

Squash baked in halves, brimming with butter & gravy, like cups you can eat

Potatoes & roots mashed up with garlic & butter, with gravy

Soup of salt cabbage, bone broth, potato, onion, with bread & butter

Roast deer, fowl, goat, with gravy

Heart cut & stewed with garlic, onion, roots

Bones of which we suck the marrow, licking our lips

Every bit of the animal used

Salt cabbage with dill & carrot, with sausage, bread, mustard

Beans stewed with fat, pepper, onion, corn, tomato

Moist cornbread with seeds to crunch, buttered & fried

Potato, mushroom, & onion fried in salt pork

Bitter, sour brews of grain, roots, fruit, honey, stored up & worried over all year

Dough fried in fat, stuffed with fish, cabbage, egg, onion, mushroom, meat, potato

Pumpkin with honey, cream, & egg, baked in a pie

New milk with the cream hovering thick on top

Nuts to crack & while away the hours

Dainty cakes dressed up pretty to behold, with sugar & cream whipped up
    in new snow

Jam of blueberry, shadberry, blackberry, mulberry, cranberry, gooseberry, cherry, grape,
    currant, apple, peach, plum, rose hip, on bread, with butter

Cups & cups of steaming, strong herbal brew, our magic potion, of chicory, of mint, of rose,
    of mullein, of wintergreen, of shelf mushroom, of any good, rich, wild thing we can find,
    with honey, with maple syrup, with cream...

Colin McMullan

May

Urbanization

Chapter 5

GROWTH OF SETTLED AREAS OF AMERICAN CITIES

MAPS SHOW EXTENT OF GROWTH AT INDICATED DATES

SAN FRANCISCO

KANSAS CITY, MO.

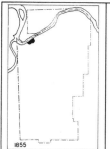   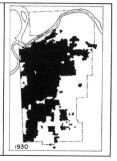

SALT LAKE CITY

DALLAS

CHARLESTON, W. VA.

EDITOR'S NOTE The consequences of urbanization. See our video on retrofitting our paved urban landscapes at ourland.tv

FEDERAL HOUSING ADMINISTRATION
DIVISION OF ECONOMICS AND STATISTICS

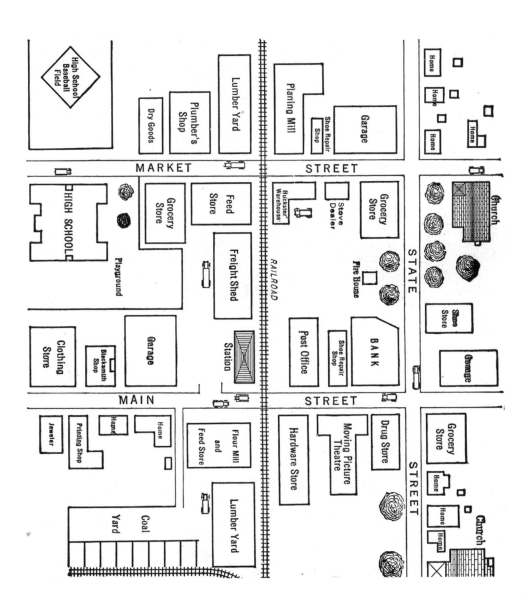

Into The Driver's Seat

Jane Bregman

My life in the city was a beautiful symphony of mechanical ignorance. Days flowed seamlessly and I never gave much thought to the inner workings of systems that facilitated many of life's conveniences. Elevators, buzzers, toasters, the Elliptical, apps, keyboards, the list goes on. I button pushed my way around until the passion of bridging the gap between field and fork brought me to the country. Once I arrived I began to grasp the value of comprehending agrarian processes and taste the empowerment this allowed. It was time for me to get into the driver's seat.

The metamorphosis started humbly enough: intro to the quad bike, which would help me navigate all three hundred acres of the farm a bit more expeditiously than on foot. I had done my fair share of snowmobiling so throttles weren't new and I had a feel for clutches from driving standard all my life. The more technical lesson came when the hay feeder was hooked on. Though I have learned all the necessary theories, I find that reversing remains a challenge. I still stop and think, "Ok, opposite of how I would have the car's wheel turned..." Fortunately the 90's vestige had no obstructions to the componentry. Being able to watch all the parts in motion allowed me to truly understand "ground driven" technology.

These were sufficient introductions to capable farm machining, but I wanted to master tools with greater damage potential. This would feel like a measure of progression. "Tractor up skilling" was next on the agenda, and I was eager. The cab

of the New Holland was filled with so many buttons and switches I felt like an astronaut. One finger amiss would surely propel the tractor into another galaxy. Once the calming hum of the engine brought me back to earth I successfully finessed the beast into all gears, wrought enough levers to power the front end loader, and pushed all the right buttons for the three point linkage bale feeder. I'm proud of learning the tractor. I believe it illustrates the passage of land stewardship from one generation to the next, and shakes up antiquated notions about what a farmer looks like.

I had turned the tractor from proverbial "big boy toy" to a vehicle of change. I was now empowered to tackle an even more fearful implement: the chainsaw. I possessed a healthy fear of dismemberment by this device my entire life. It was only with an

intensive forestry level workshop that I felt confident enough to give it a try. During the chainsaw lesson the device was stripped bare, enabling voyeuristic eyes to gaze upon the complexities of its moving parts. Proper maintenance was stressed, and safety gear became fashionable. I now look upon ear and eye protection, steel toe boots, chaps and a mitt as compulsory equipment. Using the chainsaw in the paddock became an exhilarating experience when I grasped the why behind the how.

The heroes of agrarian technology are the teachers. My journey to become a mechanically-literate female farmer has been navigated with support from these teachers—pillars of strength with enough patience, fortitude, and love to share their wealth of knowledge. Throughout this journey I've been exposed to many

mechanical systems, but the most significant one is the propagation of traditional knowledge through the art of storytelling. There is a historical snapshot within each narrative—a lesson that imparts crucial understandings for the next generation.

This is the simplicity of spoken word. While busy teaching me about engines, angles, cold starts and electric fences my teachers have given an even more important lesson: what it means to be a farmer. My life in the country is coordinated chaos. The days here are as conducted as my city days, but I see the early side of four AM more often than the late. Walking in the still dark of night along the moonlit path to the pasture, I bask in the peaceful existence of life on the land. I now know both the how and why behind my farm tools and this awareness emboldens me. I am enjoying this journey in the driver's seat and proud that all limbs and fence posts are still intact. Triumphs behind the farm gate sometimes seem insignificant, but they make rural life so sweet.

He built our cities

Vermont Food Fight

William B. Duesing

The month after Vermont governor Peter Shumlin signed into law the country's first genetically modified organism (GMO) labeling bill with a firm effective date, the Grocery Manufacturers Association (GMA), the Snack Food Association (SFA), the International Dairy Foods Association (IDFA) and the National Association of Manufacturers (NAM) sued in Federal Court to overturn. This law is scheduled to take effect in 2016; there is no trigger clause requiring other states to pass similar legislation before that happens.

With foresight, the Vermont legislature established the Vermont Food Fight Fund to help defend the GMO Labeling Law. A strong defense of Vermont's law should strengthen those in other states. You can contribute to the Vermont Food Fight Fund at www.foodfightfundvt.org.

Why are these four multibillion-dollar lobbying associations, representing the world's largest and most powerful corporations, suing to stop what the citizens want?

The GMA Board of Directors provides an indication of that organization's interests. Most of the major food corporations have a seat there. Representatives of Coca-Cola, Cargill, Nestlé, Kellogg, Con-Agra, General Mills, Kraft, Dean, and DelMonte all sit on the almost 50-person board of the organization that has sued Vermont.

These corporations have a vested interest in continuing the growth of the long-distance, highly-packaged and processed food system.

But these giants suing the state include not only the food (and junk food) industries directly affected by this legislation, but also the full spectrum of multinational corporations. NAM's 200-person board represents companies like Exxon Mobil and BP, Boeing and General Electric, Caterpillar and Archer Daniels Midland—a formidable group.

It is clear that this is an important fight for the corporations And

this fight is not just about food, it is an important skirmish in the battle for the design of our future. Corporations don't want human citizens taking control away from them.

However, unless we do take control from them and reform our political economy, we have little hope of successfully facing the challenges ahead. We must tackle an already-changed climate, the struggle to provide food for those living on Earth along with the several billion more expected to arrive in a few

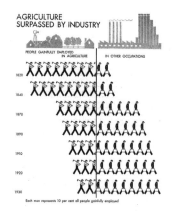

Put the Man Above the Dollar

decades, and the task of reducing our fossil fuel consumption.

If in 2050, we are feeding nine billion people with highly processed foods of marginal health benefit coming from distant sources, it may be more profitable for corporations but it is certainly not sustainable or healthy for people or the planet.

RECENT HISTORY

Vermont citizens did everything citizens in a democracy are supposed to do. They organized, educated, formed coalitions, testified at public hearings, brought in experts to educate and testify, talked to their elected representatives, even elected an organic farmer to the legislature and got their state government to support the will of the people.

This is democracy in action. But these citizens were countered at every turn by the well-funded biotech and food industry lobbying machine, and by much of the agricultural establishment.

Still the people prevailed. Food containing GMOs will be labeled.

This is not a radical law. Over nearly two decades survey after survey has shown that between 80 and 90 percent of Americans think that genetically-engineered food should be labeled. And labeling laws or outright bans on GMOs are in place in over 60 countries.

Why are so many major corporations spending time and money fighting a law that so many people want?

It is really a fight for the future of this planet's civilization. If corporations, and their narrow interest in profits from processed foods, fossil fuels, and plastic junk, are not reigned in we don't have a chance for a healthy future. More sodas and chips, double bacon cheeseburgers and fries can not feed the world.

Gus Speth's book The Bridge at the Edge of the World: Capitalism, the Environment, and Crossing from Crisis to Sustainability has the core message, in the words of one reviewer, that—"contemporary capitalism and a habitable planet cannot coexist."

The book draws on many writers and thinkers in building a strong case for major change. As one of his sources, Peter Barnes, notes "...We face a disheartening quandary here. Profit-maximizing corporations dominate our economy ...The only obvious counterweight is government, yet government is dominated by these same corporations."

Changing the most powerful economic system on the planet is not an easy feat. But Speth is clear that it is the only way we can survive.

Our current systems encourage, practically demand, that a company only grow! grow! grow!

So that is why the Vermont Food Fight is about more than food alone. It is just a warm up exercise for taking back control of our democracy. [1]

> ## "Loss and gain are brothers twain" —proverb

[1] A version of this article first appeared in CT NOFA's Gleaning enews in September 2014

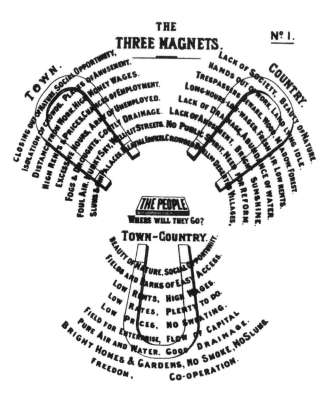

THE THREE MAGNETS.

Nº 1.

Ebenezer Howard, The Three Magnets, No. 1, 1902, in *Garden Cities of To-morrow*.

EDITORS NOTE: Howard's Garden City Movement conceptualized an ideal human habitat, where mixed-use, pastoral clusters of agriculture, light industry and residential would be optimized not for capital, but for health. Learn more about the Garden City Movement, you'll be delighted by the holism and progressive era grappling with resettlement.

FARM TENANCY

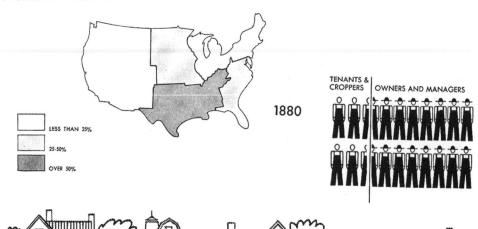

LESS THAN 25%

25-50%

OVER 50%

1880

TENANTS & CROPPERS | OWNERS AND MANAGERS

Lost City

Tony Ortega

A rectangle drawn on the chalkboard—two pairs of even-length lines, all meeting up to form a plane of space. It represented land—more specifically, farmland. It was a beginning, the first lesson of many taught to me in the old school house by the legendary Bob Cannard at Green String Farm. He was coaching us into agricultural athletes, hyping us for the wild ride that is farming. In my short time there I learned many skills which I have adapted into my life as a farmer. By trimming down the first step to a four sided shape, Bob gave me courage to simply start.

Two years later I was standing in the middle of my own bare rectangular acre—land! This acre—neighbored by weekly motels and high rise casinos—sits in downtown Reno, Nevada, a high desert city in the vast Great Basin. For over a decade the land was overlooked and vacant, but to our farm it is a gem. Many people do not think of Nevada as a food producing state. With 300 days of sunshine and a meager seven inches of rain per year, it is not easy to cultivate in Nevada's dry and windy climate. Here, however, the soil is fertile and the Truckee River flows right through town. We do grow food in this place, our urban acre is testament to that. So with eager dreams and a flood of ideas we threw ourselves into creating what is now Lost City Farm.

We would have to wait a year to touch the soil. Our first steps were negotiations with city planners, city council, and lawyers as we ran into antiquated zoning regulations. We knew that the road was rough, we faced every obstruction in the way. Our courage was driven by the desire to feed the community in which we live. After months of working to create better land use regulations, the Urban Agriculture Ordinance was adopted in Reno. Urban farming would now be possible in the Biggest Little City.

The year we broke ground was intense and exciting. We put energy back into the soil, created infrastructure and a name and niche, and most importantly we began to sow seeds. We were finally growing food.

Now in our second season, there are days when it's difficult to remember our reasons for starting this journey. The farm is fertile. The acre is green. The "like" buttons have been clicked thousands of times, but with the overwhelming to-do lists and a ceaseless exhaustion during midfarm season, much of our enthusiasm is squelched. I sometimes question the life I have chosen—working two additional jobs while learning my way around a strange piece of land in a city not accustomed to edible green space. I feel defeated when the produce doesn't sell out but our Farm T-shirts have become a fashion trend. When 25 extra bunches of carrots become more of a burden than a bounty, I can't help but wonder why it's so hard to feed people.

Farming is like ritual, tied up with so much history, method and theory. To become a farmer and to start a farm we accept the responsibility carried by all those farmers before us. I feel this ancestry watching over me, and it is comforting and lonely all at once. The number of new young farmers in America is growing, but we are spread out across a vast landscape, making our community seem sparse.

The work will continue to be done, because as farmers we know the value of that work. But the duty does not rest solely on the farmer. It is a symbiotic relationship between farmer and community. Of course not everyone wants to farm—it is hard work—but we all need to eat. We need to start putting our money where our mouths are, become responsible consumers, and take time to truly comprehend our food systems. We must recognize that these are actions and we must support young farmers through our behavior so that we can all thrive financially and emotionally.

Our little acre has become a place of importance and nurturing work. It is a space that is visible again—a

Jenny Edwards

noticeable mark on the map. We continue to sow seeds and do our best to help them along, looking forward to a time when we will outgrow novelty and become the neighborhood produce stop, the summer hangout, a meeting place, the Farm.

Bob once said that the toughest challenge is learning to execute your commitment. Honoring that commitment has proven larger than the four sides that delineate Lost City Farm. It is the land we cultivate, the location where we learn the economy of our movements and the fruits of our labor. It is a place to start. But in order for it to continue, our acre must be thriving all around the community–as food on people's tables, as a dialogue with city government, and as a resource for education.

We have no quick solutions, just time. Slowly and steadily we cultivate our acre of land with seeds, soil, and water. Slowly and steadily we hope to inspire others to cultivate the place we live with fresh ideas, active participation, and patronage of small farms.

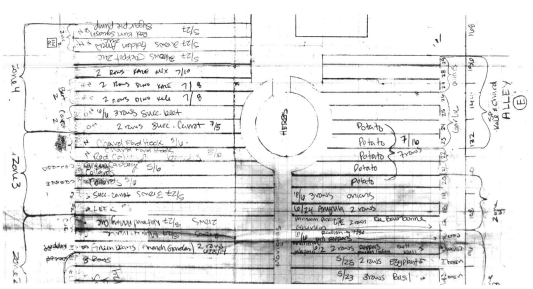

The Lost City farm plan used year to year to keep track of planting.

Fruit and Nut Trees

Apple

Oregon's, and the nation's, leading fruit tree is the apple. Earliest recorded history mentions it as the "gift of the gods." The apple is most easily identified by the characteristic round fruit known as a pome, a fleshy fruit having seed borne within cells, or carpels, at the core. There are many species of apples and several thousand varieties. Apple leaves are oval, mostly pointed at tip and rounded at base, soft and dull. The large, showy flowers are borne in clusters of five or six in each group. Some apples are called "crabs"; these are usually small in size but have attractive flowers and spicy fruits.

Apple seeds are spread by animals and birds so that trees frequently escape to fence rows, abandoned fields, and even cutover forest land. Fruits of these so-called "seedling apples" show much variation in size, flavor, and color.

Pear

The pear is another traveler like the apple, having come to us from western Asia and China. Pear trees are strong and upright, sometimes being 50 or more feet high. The leaves are oblong, borne on short spurs, hard in texure with prominent veins, and have a bright green color. Pear fruit varies in shape from round to oblong, but, like the apple and quince, it is a pome type having an inner core containing brown seeds. The white flowers appear in dense clusters with 4 to 12 flowers in each. Over a thousand varieties have been named but only a half dozen are grown commercially. Oregon is a leading producer of pears.

Quince

The "true" quince which originated in Asia is prized for its fruit and flowers. This species grows as a poorly formed tree or shrub 15 to 30 feet high. The large flowers are white or tinged with pink. The globular or pear-shaped fruit is borne at the end of a twig and may reach a diameter of 4 inches. The fruit is straw-yellow when ripe, very firm, fuzzy, and highly fragrant. The fruit is sour to taste, but its flavor is excellent in are devoured by deer and other wildlife.

Cherry

Cultivated varieties of sweet and sour cherries are kin to the wild chokecherry and to the laurels. The trees are tall and erect with reddish brown bark which peels off in rings. Cherry flowers are showy. They are creamy white and borne on long stems in dense clusters of 4 to 8 blossoms. Fruits may be red, yellow, or black, heart shaped or pointed. Cherries, like peaches, apricots, and plums, are known as stone fruits with the seed enclosed in a hard stony shell.

Peach

The peach is one of our introduced fruit "cousins" from China. It made its way to America in the 16th century. The tree is twiggy, with slender limbs, and usually bears one to three buds at a node or joint. The pink flowers are solitary, and appear before the leaves. Peach trees are shorter (generally under 20 feet in height) and more round than apples and pears. Dozens of varieties are known, including red, yellow, and white-fleshed types, with some having purple leaves and double flowers that are planted as ornamentals. An interesting close relative is the smooth type of peach, called a nectarine.

Plum

The many types of plums come from three continents, Europe, Asia, and America, and differ in characteristics of the buds, flowers, and fruits. Our best known plum is the Italian prune. Plums have attractive white flowers borne in dense clusters on flower stems about one inch long. Fruits come in a variety of colors from almost black through shades of red, purple, blue, green, yellow, and white. Our native American plums (some in Lake and Klamath counties) are small and tart.

Apricot

The trees are somewhat similar to almonds, easily distinguished by large, broad leaves which are sharp pointed; and round, bright yellow fruits which ripen in July. The flowers are $\frac{3}{4}$ to 1 inch across, pinkish to nearly white, and appear very early. The flesh of the fruit breaks free from the inner stone which is flattened and smooth.

Pick Yer Own: Building Community Through DIY Harvesting

PB Floyd (Jesse Palmer)

Urban harvesting has numerous positive aspects: it nurtures community and encourages talking to your neighbors, it promotes consumption of locally grown, non-fossil fuel tainted food, it is do-it-yourself (DIY) so you learn new skills, it gives you a valuable connection to the earth and its natural cycles which people in cities often lack, and it permits you to experiment with distribution outside of the market system.

Harvesting

It is hard to believe how much fruit one small tree can produce! The first step is identifying fruit trees near your house. In our neighborhood, there are many fruit trees that are not harvested because the people living in the house with the tree don't do the work. You can walk around and make a map in your head or on paper when the fruit is ripe and note which trees seem to get harvested and which don't.

Then comes the exciting, but perhaps uncomfortable part: you have to talk to your neighbors and ask if you can harvest their tree. We left a note with our phone number or visited if we already knew the neighbor. It seems that neighbors talk to each other less and less in the modern world, and that's too bad. Perhaps it is the rise of internet and car culture—a culture of isolation and loneliness. When I was growing up, I knew people maybe within a block or two of my parent's house. Since then, I've sometimes lived somewhere and not even known the person next door! Meeting neighbors moves the idea of "building community" from just a slogan to reality. Communities where people know each other can organize to resist hierarchical power structures and build voluntary, non-market-based alternatives.

When our house asked to pick our neighbor's trees, they always said yes—sometimes with great excitement. The neighbors were usually happy to have someone use the food and picking fruit trees avoids a rotting mess when the fruit falls to the ground.

The real fun begins once the fruit is picked. The first thing you learn is that fruit ripens all at once. So harvesting isn't like going to a grocery store and only getting what you need at that time. When you harvest, you either have nothing, or way too much of a particular thing. Our ancestors knew what foods were in season at what times like the back of their hands, but in a world with fruit shipped around by airplane, we get fooled into eating like the seasons don't exist.

Once you start to notice what is in season in your area, you may begin to adjust what you eat and seek out locally grown food in season. Eating like this drastically decreases the amount of fossil fuels required to keep you fed. Noticing these things adds richness and connection to your life experience just as living a mechanical life disconnected from the earth and its cycles can strip meaning away.

Preserving and distributing

Preserving foods opens lots of DIY opportunities. You could dry huge quantities, you could make juice and ferment it into hard cider, you could can fruits and vegetables, or you could just give all the excess away!

This is another opportunity for building community and developing alternatives to the market-based distribution systems that exist under capitalism, as well as an opportunity to consume fossil fuel free food. These days, you can get organic and fair trade food, but it is almost impossible to get fossil fuel free food! Figuring out how to grow, distribute and eat fossil fuel free food is the next frontier, because when it comes down to it, burning fossil fuels is killing us. Organic goes part of the way, of course, since a main ingredient in conventionally farmed food is chemical fertilizers, which cannot be made without fossil fuels. But eating organic avocados imported from Chile in January misses the point of "organic"— eating now shouldn't destroy the environment's ability to grow food for our grandchildren.

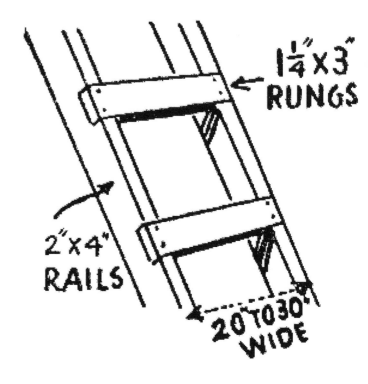

Epilogue

It would be easy for cities to plant many more urban fruit trees to supply local food needs, except, naturally, for the law. Most cities make it illegal to plant fruit trees on the parking strip—the little strip of useless grass between the sidewalk and the street on millions of miles of urban streets. The idea behind the law is that urban fruit trees would be messy—the assumption is that no one would pick the fruit and that it would thus fall to the ground and rot.

These laws are stupid. Why are modern people so afraid of messy things? Life is messy from birth to death and decay—get used to it! A few of the trees we harvested were "illegal" fruit trees on the parking strip. This spring, we're going to plant a few "illegal" fruit trees on our parking strip. We're likely to "get away with it" since we're planning to harvest them and keep the area clean. What if millions of people planted urban trees on parking strips and other unused land? Or better yet, what if the silly laws were eliminated and cities planted fruit trees on all available parking strips, perhaps with the formation of neighborhood harvest committees or by hiring local youth over the summer to tend, harvest and distribute the fruit?

The New Spirit of the New Agriculture

By HERBERT MYRICK

The girl's face is to the future;
She knows not why, but Nature knows.
Looking o'er the bounteous fruit —
Whose richest bloom is not more rare
Than her own unsullied beauty —
She feels a strange joy and gladness,
The hope and confidence of youth.
    Self-reverence practicing,
    She grows in power and poise,
    Composed, resourceful, strong,
    In character nobly planned,.
    By wise training truly fit,
    To do well her part in life —
THE GLORY OF AMERICA!

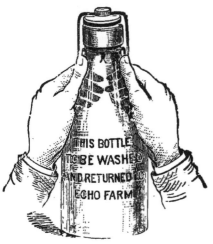

THE FARMERS ADVOCATE

Volume 27. No. 47   TOPEKA, KANSAS.   June 8, 1905.

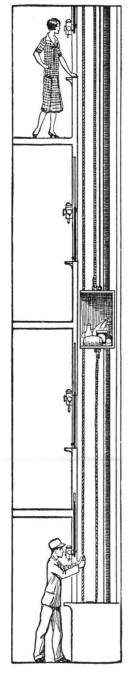

FARMERS SHOULD READ.

The lawyer or the doctor is supposed to be well read in the literature of his profession. There is no reason why the farmer should not seek to be as well read along the line of his work. His reading possibilities are now very large. This is at should be. No other occupation brings one in closer touch with so many of the profoundest facts of life. No man can hope to apprehend these, or to be successful even in one branch of agriculture, without seeking to incorporate into his own the results of the work of others.

Progress in agriculture is not so much the result of what a few men of genius have done as it is the result of the patient work of thousands of toilers, who have given to it the skill of their hands and the best thought of their minds. The farmer of today can be all that these thousands of fellow workers have helped to make him. But it is only through the reading and the study of the literature of his occupation that he can realize these possibilities.

But do we go far enough when we advise farmers to read simply as farmers? Is not the man more than the farmer? Certain it is that a man is always more than his work. No man should make farming the end of life, but should rather minister to life. The growing of the largest crops and the breeding of the best animals is not so important as the development of himself in character and manhood. Not all of a farmer's reading should be of an agricultural nature, but some of it should be of such a kind as to develop himself as a man, to increase his general knowledge, to broaden his view of life and to give him wider sympathies.

"He was born a man and died a grocer," was said of one. Something of the kind might be said of some farmers. The farmer who gives some attention to his own mental improvement will be more rather than less successful in his work. Farming is a matter of brain as well as of the hand. The more he trains his mind to think the better he will be able to work out for himself some of the problems of present day agriculture. No one can learn to think except by thinking. And in order to think, one must furnish his mind with the material for thought. This will be gathered by reading and study. "I think thy thoughts after thee," said a wise philosopher of old. So in thinking over the thoughts of others and making them his own, will one learn to think.

More than this, the farmer is not only an individual, who is to develop in himself the possibilities of the race, but he is also a very important factor in our economic and social life.. As such he cannot dodge the responsibility which is laid upon him. He is looked to as contributing a degree of stability to our political life. This means an opportunity to make his influence felt, which should not be neglected. More than most men, he is concerned in the political and economic questions of our country. He should seek to help settle these in a way, which will help both himself and the whole country.

For the necessary training and knowledge to help him in this work he needs to read. He must keep in touch with the most important events of the day, and understand the history and government of our country. All this will not involve any great outlay of money or an amount of time beyond his control.

If these suggestions were carried out by farmers generally, they would not only become better farmers, better citizens and better men, but life to them would have a larger meaning, and rural communities would be the more worth living in. And more than this, their farms and property would increase in value, for values rest not in real estate and things, but in the character and condition of the people among whom they obtain.—American Agriculturist.

The Fascinating History Of Italian Radio

Pete Tridish

Italy happens to be home to one of the largest FM transmitter industries in the world. A lot of transmitters branded as being "Made in the USA" are actually made with components from Italy. The reason is interesting.

When the Communists came to power in the mid seventies, they managed to appoint a communications ministry that had anarchist tendencies. The ministry abolished all radio and TV licensing, and a fascinating free-for-all ensued. Every labor union, student group, community center, church, and two bit capitalist set up their own radio station—one famous one was called simply " Radio Alice." It was four beautiful, innovative years of radio anarchy.

However, there are only so many viable FM frequencies on the band—most cities can fit about 40 or 50 depending how tightly they are packed. Instead of competing for licenses from the government, everyone competed by using larger transmitters to overpower their competitors.

Obviously, whoever had the most money won these competitions. From this came the rise of Silvio Berlusconi, who was able to build a largely-unregulated media empire and use it to accumulate even more money, power, and influence. This history of Italian media is an interesting parable about how anarchism done wrong can turn into a stupid kind of libertarianism. It is not enough to get rid of state limits; capital must also come under community control. In this historical moment, even a regulatory state is usually not strong enough to contain a force as powerful as capitalism, which will use its economic power to shape society for its own interests.

A byproduct of all this was that Italy ended up with an unusually high demand for transmitters. With so much money coming in, the industry had both the capital and human resources necessary to become the world's largest exporter of transmitters. The unregulated market was great for building the transmitter industry. The Italian people however, who had to suffer under Berlusconi for so many years, got the raw end of the deal.

Philosophy For May

Nate Chisholm

There is a long tradition of Almanacs being a pantry for agricultural philosophies. But it's May, and I don't have time for philosophy in May.

In January there was time. In the winter, storms off the Pacific turned our voluptuous hills green as Ireland. Then, it was easy to know what the grass wanted—just to be left alone as long as possible. With thirty-degree nights, the grasses grew slow even with all the rain, and they kept growing even when the cattle plucked their leaves.

But in May, after the rains have stopped, its trickier—it's hot. We can see the summer fog, poised on our western horizon, but we can't feel it yet. This time of year the grass seeds are green and the grass stems are resilient. The plants might not regrow a leaf when grazed.

The grass wants something, but we don't know what...

Around the solstice the seeds will ripen and shatter. Then my job will become simple again. Having dropped their cargo the grass stems will give up easily, turning to dust as soon as cattle touch them. Having done their job, the only thing left is to get out of the way. Get out of the way of the little seedlings that will be coming up after the next Pacific storm. Get out of the way of the leaves that will cover all the mistakes we made this season.

But listen to me ramble on, I better get back to work. Hopefully there will be more time for philosophy in June. Or maybe that isn't necessary. Maybe this is all the agricultural philosophy we need. Working to find out what the grass wants, we restore habitat for savanna sparrows and checker lilies. And we restore human habitat too. Human habitat is better than any philosophy—better than any agriculture, even.

Willa Mamet *Parts of the Cow*

ANOTHER "STAMP" CITY

Food Distribution Plan to Seattle, Rochester and Dayton

Birmingham, Ala, has been selected as the fourth city in which the food order stamp plan for distribution surplus food products will be started, according to an announcement made July 11 by Secretary of Agriculture Henry A. Wallace.

Seattle, Washington, has been named by the department of Agriculture and the Federal Surplus Commodities Corporation as the third of the six experimental cities to test the "Flood Stamp Plan" for distribution of surplus commodities through regular channels.

The first city chosen and announced on April 18, was Rochester, New York. The second city selected was Dayton, Ohio, and the plan has been in operation in Dayton since June 6.

Seattle, selected as the third city, will soon be operating under the new stamp plan which provides books of stamps—both orange and blue—for WPA workers and persons on general relief, which enables them to receive a more balanced diet, as well as additional food for their families.

The orange stamps are for the purchase of any food desired, while the blue stamp—in the amount of 50 per cent of the orange stamp—are for the purchase of those commodities which the department of agriculture designates as being in surplus.

EDITOR'S NOTE In the Great Boston Molasses Disaster of 1919, a molasses storage tank burst in the summer heat and a wave of molasses swept through the streets at around 40 mph. 21 people were killed, 150 were injured. The molasses was being used to make rum, and a whole new crop of small-scale distilleries has popped up in contemporary Boston. Shipping booze is an excellent way to make Sail Freight more profitable, so if anyone is in touch with this industry then send them our way!

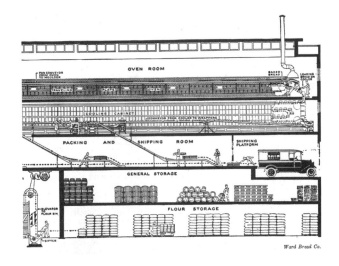

Ward Bread Co.

I do not agree with the conclusions recently reached and expressed by one of our greatest producers and financiers that the farming of the future, to be profitable, must be on a large scale; that it must be so systematized and such labor saving be used that, at the touch of the master hand, both the mechanical and human machines will move forward in perfect co-ordination. These are the logical and natural conclusions of the materialist who thinks mainly in terms of dollars and cents, but we are taught by a far greater authority that "life is more than meat and bread.". Therefore, I want to go on record here and now that the dollar, essential as it is—the man himself is far more essential, that man should own the dollar and not the dollar own the man, that the nation controlled by good citizens is far safer than a nation controlled by great material wealth, I will go still farther and say that a country in which farmers—small farmers who own and cultivate their own land—small through the acreage may be—is the safest country or nation on this earth.

June

Logic & Context

Chapter 6

Farming Pregnant

Janna Berger Siller

Here I am, slicing pigweed roots out of the onion beds with a hoe while another human grows in my body. And there I am, sinking a thousand pounds of steel into the earth with the flick of a finger on the tractor's hydraulics, tilling in the spinach to make way for sweet corn while ears form and eyes take shape under my belly button. Here I am, teaching a group of mosquito-dodging teenagers how to identify the sweetest sugar snap pea while my blood volume increases steadily and my bladder squishes itself into an ever-narrowing space. And there I am, loading crates of lettuce while carrying another body—separate but enmeshed—back and forth with me from the cold shed to the truck bound for market.

The images on pregnancy blogs are sketched in soft pastels but my experience is sweaty and surreal and the colors of the world look bolder than ever. Who has bent over fertile soil, back muscles gripping between spine and ribs while the firm perimeter of their uterus held close its bulge? Who has leaned over their big belly, kicked from the inside, to put seeds in the ground—both creator and container of new life?

I know that the world is dangerously overpopulated and devastatingly deforested. Yet I find myself striding purposefully across farm fields in the heavy footsteps of pregnant humanity. I suppose that I am producing a new human for the same vague reasons that I spend my days producing food for the humans who are already here. Despite my fear of our species' capacity to hog resources and cause suffering, I choose to follow my impulse to live rather than dwell in my shame for the exploitation involved in living. I am drawn to the intimacy of what it means to be an animal on earth, a human between the soil and the sun.

And so here I find myself picking peppers and washing radishes and planning out next year's crop rotation while singing to the creature growing in my abdomen. My heart pumps, fueling my work, and so does her heart within me.

Chestnut the Future

James Most

WHY CHESTNUTS?

A fresh roasted chestnut is incomparable to any other nut—like a chunk of caramelly sweet potato. They contain very little fat and have a carbohydrate content similar to wheat or rice, hence their reputation as the "bread tree." They can be dried and milled into flour to make cakes, breads, and pasta. Animals also go nuts for chestnuts—for thousands of years, hogs, horses, and sheep have been sustained and fattened on chestnuts.

The market for fresh chestnuts in the United States is currently insatiable. Only about 5% of the chestnuts consumed in the US are produced here—the rest are imported from Asia or Europe. Most growers will sell as many chestnuts as they can pick up—many even establish waitlists for their customers.

WHAT CHESTNUTS?

Relatives of beech and oak, there are four main chestnut species of the Castanea genus worldwide—The European (Castanea sativa), The Chinese (Castanea mollissima), The Japanese (Castanea crenata), and The American (Castanea dentata). These are not to be confused with the unrelated but similar looking Horse chestnut (Aesculus hippocastanum), which is mildly poisonous.

Since virtually all of the American chestnuts of the Eastern forests were wiped out by a chestnut fungal blight in the early 1900s, North Americans primarily grow European, Chinese, and European X Japanese hybrid (EXJ) chestnuts. American chestnuts are most susceptible to the blight, followed respectively by the European and EXJ varieties, and finally the Chinese variety which is almost immune since the blight is indigenous to Asia.

The blight still lingers east of the Rockies, so most chestnut growers in the East grow Chinese varieties. In the Midwest however, some EXJ growers are experimenting with a viral inoculation that reduces the damage of the blight. Additionally there are ongoing attempts to cross American and Chinese varieties to create an American chestnut with the blight resistance of a Chinese tree.

West of the Rockies, chestnut growers can freely grow European and European X Japanese hybrids without concern for the blight. The blight is present in the West, but climatic conditions are not conducive to its survival. If the blight does establish in the future, European X Japanese hybrids are likely to be the most compatible with current blight battling strategies.

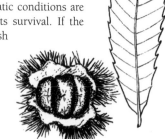

Chestnut folage and nuts

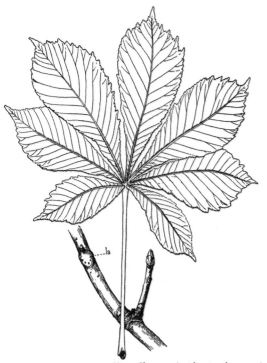

Horse Chestnut. Avoid eating these ones!

HOW CHESTNUTS?

Chestnuts are a hardy, long lived, and drought resistant tree that can handle rocky and steep soils. Despite their resiliency, it is easy to kill chestnut trees put in the wrong situation. Heavy, clayey, and overly wet soils spell doom for chestnuts, so plant chestnut trees in the location with the best possible drainage.

A planted chestnut seed will produce a tree, however seedling chestnuts can vary widely in size, peelability, and flowering times and make it difficult to plan an orchard. Whether you are growing European, Chinese, or EXJ hybrids, there are hundreds of selected cultivars to choose from. Planting grafted selected varieties or grafting selected varieties onto seedlings in the field ensures chestnut size and quality in your orchard. Seedling trees can also take many years to come into production while grafted trees can begin producing in 2–3 years.

Chestnut pollination is finicky. They are wind-pollinated and self-sterile, so it is best to plant chestnuts like you would plant corn—in larger groupings or blocks. Similarly, knowing your varieties by planting grafted trees allows you to align flowering times to ensure proper pollination. Trees should not be planted too close together—cramped trees are less productive, more disease susceptible, and require much more pruning and maintenance. Putting extra energy into mulch, fertility management, and consistent irrigation pays off as well-maintained trees will come into production quite quickly.

WHOSE CHESTNUTS?

A baby tree started from seed can reach plantable size in 2 seasons, but grafting can be tricky. Chestnut varieties have different "bloodlines", giving rise to graft incompatibility. To ensure grafting success, scion should always be grafted onto seedlings of the same species, or even from the parent directly to the child if possible. (i.e. Betizac scionwood onto Betizac seedling). Finding grafted chestnut trees may be difficult, but the information is out there so remember to talk to local growers.

Progress of the Chestnut Project

In an effort to promote chestnuts as a crop on the West Coast, the Chestnut Proliferation Project will be distributing grafted European X Japanese hybrid chestnut trees to farmers, ranchers, and other landowners who are keen to diversify their agriculture with chestnuts as a perennial tree crop. Funded by a grant from Nutiva, these trees will be distributed free or subsidized along with consultation and establishment support.

WHAT

Hundreds of baby chestnut seedlings have germinated and are putting on lush growth in nursery beds and pots on Orcas Island, Washington. These young trees are being propagated from seed, and will spend all of 2014 growing and getting thick enough to be grafted to selected varieties in the spring of 2015. Starting in 2016, these trees will be ready to transplant out of the nursery and into orchards.

WHY

The goal of this project is to encourage landowners, farmers, and ranchers to incorporate functional tree crops into their land management practices. Tree crops benefit all elements of land management. From an environmental angle they control erosion and carbon uptake, and from a business angle they encourage crop diversification and resiliency and reduction of dependence on grain based animal feeds. Sponsored by a grant from Nutiva, these chestnut trees will be distributed to ranchers, farmers, and landowners on the West Coast with consultation and establishment support.

WHO

With fiscal support from Nutiva and project sponsorship from the Greenhorns, James Most and Sara Joy Palmer are leasing a site on Orcas Island for the chestnut nursery and are currently propagating hundred of trees. In addition to propagating trees, James and Sara Joy have begun reaching out to potential landowners on the West Coast to site future chestnut orchards.[1]

---

1   To find out more: Contact James at jamesmost@gmail.com to find out more about the progress of the project and to talk all things chestnuts.

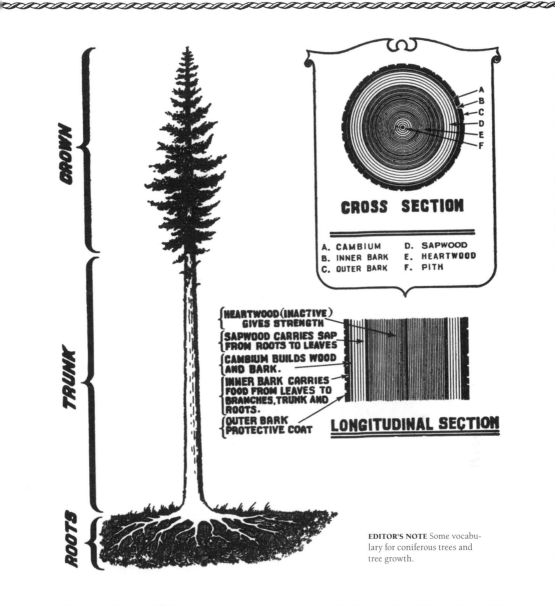

CROWN

TRUNK

ROOTS

CROSS SECTION

A. CAMBIUM   D. SAPWOOD
B. INNER BARK   E. HEARTWOOD
C. OUTER BARK   F. PITH

HEARTWOOD (INACTIVE) GIVES STRENGTH

SAPWOOD CARRIES SAP FROM ROOTS TO LEAVES

CAMBIUM BUILDS WOOD AND BARK.

INNER BARK CARRIES FOOD FROM LEAVES TO BRANCHES, TRUNK AND ROOTS.

OUTER BARK PROTECTIVE COAT

LONGITUDINAL SECTION

EDITOR'S NOTE Some vocabulary for coniferous trees and tree growth.

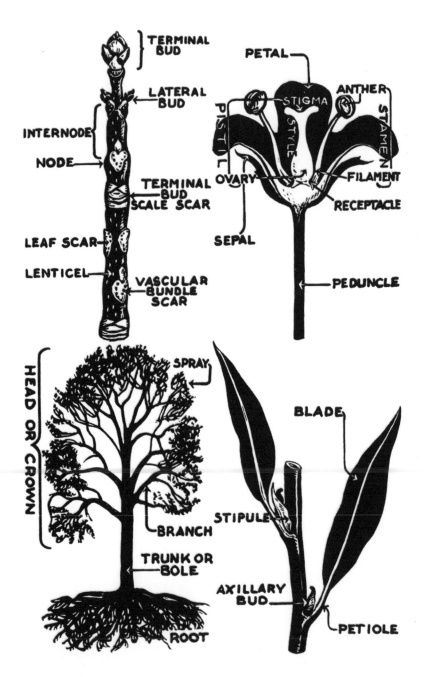

TERMINAL BUD

LATERAL BUD

INTERNODE

NODE

TERMINAL BUD SCALE SCAR

LEAF SCAR

LENTICEL

VASCULAR BUNDLE SCAR

PETAL

STIGMA

ANTHER

PISTIL

STYLE

STAMEN

OVARY

FILAMENT

RECEPTACLE

SEPAL

PEDUNCLE

HEAD OR CROWN

SPRAY

BLADE

STIPULE

BRANCH

TRUNK OR BOLE

AXILLARY BUD

PETIOLE

ROOT

EDITOR'S NOTE Some vocabulary for deciduous trees and tree growth.

Tree Terms

PARTS OF A LEAF

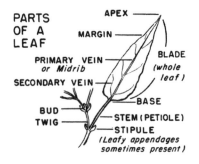

APEX

MARGIN

BLADE *(whole leaf)*

PRIMARY VEIN or *Midrib*

SECONDARY VEIN

BASE

BUD

TWIG

STEM (PETIOLE)

STIPULE *(Leafy appendages sometimes present)*

BRANCHING

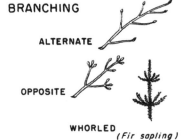

ALTERNATE

OPPOSITE

WHORLED *(Fir sapling)*

NEEDLE-LIKE or CONIFEROUS LEAVES

In **BUNDLES** *(Pines)*   **SCALELIKE** *(Juniper)*   **SINGLE** *(Douglas fir)*

BROADLEAF or HARDWOOD LEAVES

SIMPLE *(most broadleafs)*   **COMPOUND PINNATE** *(Ash)*   **COMPOUND PALMATE** *(Horsechestnut)*

LEAF SHAPES

Lance-shaped or **LANCEOLATE** *(Willow)*

Egg-shaped or **OVATE** *(Cottonwood)*

Reverse Egg-shaped or **OBOVATE** *(Chokecherry)*

ELLIPTICAL *(Cascara)*

Heart-shaped or **CORDATE** *(Aspen)*

LEAF MARGINS

ENTIRE
(Oregon Myrtle)

Wavy, or
SINUATE
(Pacific Dogwood)

Sawtoothed,
or **SERRATE**
(Elders)

Doubly - sawtoothed
or **DOUBLY SERRATE**
(Thinleaf Alder)

LOBED
(California
Black Oak)

LEAF POINTS (APEXES)

Sharp
or **ACUTE**

Drawn out
to fine point
or **ACUMINATE**

ROUND

Abruptly
sharp - pointed
or **MUCRONATE**

Blunt - pointed
or **OBTUSE**

LEAF BASES

Wedge- shaped
or sharp
CUNEATE or ACUTE

Uneven or
ASYMMETRICAL

Heart - shaped
or **CORDATE**

Blunt - pointed
or **OBTUSE**

Flattened
or square
or **TRUNCATE**

Remember, Leaves From the Same Tree Differ!

Leaves from the same tree species will vary in size and shape. With practice you can soon learn to recognize these variations. Leaves of the black cottonwood, for example, may be egg-shaped (ovate) or pear-shaped, and either heart-shaped (cordate) or rounded at the base. Their margins range from wavy (sinuate) to scalloped (crenate) and finely sawtoothed (serrate). Besides the terms pictured here, botanists use additional ones to describe shapes and margins of leaves.

BULLETIN 132

Why Massachusetts Needs Town Forests

Premium Plantings

FITCHBURG'S TOWN FOREST
This is said to be the first legally established town forest in the United
States.   White pine about forty years old

MASSACHUSETTS FORESTRY ASSOCIATION,
4 Joy Street, Boston, Mass.

WAYS AND MEANS OF CREATING A TOWN FOREST
Some Suggestions as to How to Go About It

1. Have a committee of representative citizens investigate the possibilities of the town for town forests and put an article in the town warrant, providing an appropriation for the purchase of the lands recommended by the committee, or authorizing the sale of bonds for the purpose.

2. Have an organization like the Chamber of Commerce, the Women's Club, the Village Improvement Association, the Boy Scouts and Girl Scouts or any other civic organizations take the lead or co-operate in presenting the matter to the voters at the town meeting.

3. Have any one or a group of the organizations raise the money for acquisition by public subscription, buy the land and present it to the town for a town forest.

4. Have the water supply area converted into a town forest.

5. Make the forest land of the Poor Farm into a town forest.

6. Have the committee solicit gifts of land for the purpose.

7. Have any unused public land fit for growing timber converted into a town forest.

8. Extend the present parks by adding land for town forests.

9. Urge some individuals to make a gift of a memorial forest.

10. Create a forest as a memorial to the soldiers of the town

11. As a means of raising money ask individuals and business firms to subscribe enough money to pay for one or more acres of the land to be bought.

12. Do not overlook the argument that the State will furnish young trees as soon as they are available, free of charge, for the planting of the forest.

13. Remember also that the management of the town forest can be turned over to the State Forester and in that way the town will receive technical advice free of charge.

14. An inspection of the assessors' books will enable you to locate readily the possible areas that are fitted for town forests. The assessors will no doubt be able and willing to furnish information about any specific tracts.

15. The State Forester is available to give you technical advice as to the fitness of any tract or tracts for town forests.

16. The offer of the Massachusetts Forestry Association to plant 5000 trees in the town forest should be of assistance in arousing interest. It will insure a beginning in the creation of a profitable forest.

Farmers and Wildlife

Megan Prelinger

Farmers are partners in the land with wild animals. The diverse farm ecology that benefits food production also benefits the entire web of life. Farmers who encourage crop diversity, wind breaks, and patches of wild space on their land are rewarded with a land rich in fellow animal species. Many animals are beneficial to farming, helping to control pests and weeds, pollinating crops, and enriching the soil.

The information offered here is aimed at the situations when a farmer encounters an animal in distress. Everyday life in the wild is dangerous for animals, and no one expects farmers (or anyone) to change that. However, many threats to animals are directly or indirectly caused by people. The advice on this page is directed at reducing the impact on wild animals of treatable and avoidable situations.

Animals can fly into power lines, become entangled in wire or equipment, sustain gunshot wounds, become entangled in fishing line and tackle, and become poisoned by non-food substances. They can also be hit by cars, boats, kites, and tractors, or hurt by dogs, cats, or uninformed people. Animals sometimes also nest right in the middle of a field that needs to be plowed. Wildlife rehabilitators have developed their healing arts partly in response to the ethical problems posed by the impact of people on wildlife. As a farmer, you can be ready to deal with these situations by knowing who the wildlife care professionals are in your area and being able to reach them.

Here is how to educate yourself: Find out who the wildlife care professionals are in your area. There may be a wildlife clinic listed online near you, or your local Animal Care and Control agency will be able to tell you. A local veterinarian or U.S. Fish and Wildlife officer, or even the police or sheriff may also be able to help you locate these resources. Wildlife rehabilitators are licensed by U.S. Fish and Wildlife (who govern the laws around capturing, handling, and housing wildlife). There is a National Association of Wildlife Rehabilitators: www.nwrawildlife.org. Many rehabbers work out of their homes, so don't be surprised if the nearest licensed rehabber turns out to be your neighbor. It's best to figure out where to take an injured animal before the emergency happens, rather than after that Golden Eagle has caught its foot in your tractor wheel and started going hyperthermic. Have the phone number of your local wildlife rehabilitator(s) on your emergency telephone list.

To be even more prepared, call the local wildlife rehabilitator when you have a quiet moment, and find out what their specialties are. Some wildlife centers treat all animals, others specialize in birds or in mammals. Off-season, check out their facility.

Read the Guidelines for Safe Handling of Wildlife so you are prepared for emergencies.

It is important to be aware that many laws and regulations govern the handling and capture of wildlife, and that these vary state by state. Look up the phone numbers of the state and federal wildlife agencies in your area, and put those numbers on your phone list as well.

For example, if a flock of White-Faced Ibises has nested in your field, and you need to plow it, there are laws to protect wildlife that govern what you can do about it. Your local U.S. Fish and Wildlife officer can advise you, and can possibly assist you in getting a permit to relocate the nests. If your farm is a haven for endangered species, educate yourself about those species and about any special considerations that may apply.

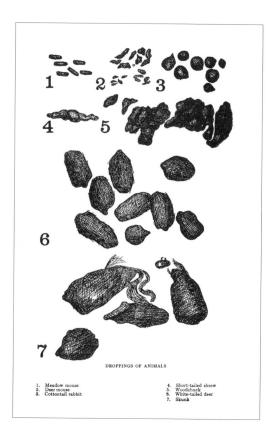

DROPPINGS OF ANIMALS

1. Meadow mouse
2. Deer mouse
3. Cottontail rabbit
4. Short-tailed shrew
5. Woodchuck
6. White-tailed deer
7. Skunk

For further information:

The National Wildlife Rehabilitators Association: www.nwrawildlife.org
U.S. Fish and Wildlife Service home page:

Write in the contact information for your local wildlife resources and specialists here:

Local Wildlife Rehabilitator(s): _____

Local U.S. Fish and Wildlife Office: _____

Local Animal Care and Control: _____

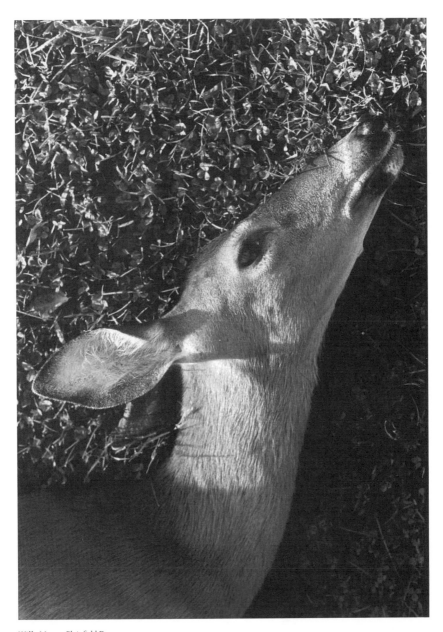

Willa Mamet *Plainfield Deer*

WATER GOING FROM SEA TO LAND.

LESSON 10.

Drainage.

Do the clouds drop upon the earth more water than is needed to moisten the ground?

They do.

What becomes of the water that is not needed?

It sinks down through the earthy matter above the rock, and into the cracks of the rock.

Does it remain there?

No; it works along under the ground, and at length comes out through little openings called springs.

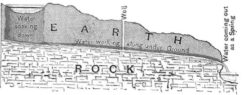

What does it then do?

It flows in the form of rivulets down the slopes of the land.

Where do the rivulets lead it?

They lead it into larger streams called brooks. Many brooks unite until the stream they form is large enough to be called a river.

What is a river?

A river is a large stream of water flowing through the land.

What is a lake?

A lake is a hollow place in the land filled with water from a stream or from springs at the bottom.

Which way does water always run?

Down hill.

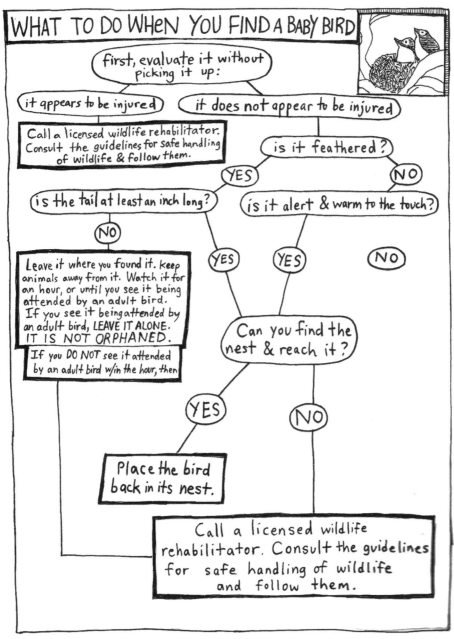

WHAT TO DO WHEN YOU FIND A BABY BIRD

first, evaluate it without picking it up:

it appears to be injured

it does not appear to be injured

Call a licensed wildlife rehabilitator. Consult the guidelines for safe handling of wildlife & follow them.

is it feathered?

YES — NO

is the tail at least an inch long?

is it alert & warm to the touch?

NO

YES   YES   NO

Leave it where you found it. keep animals away from it. Watch it for an hour, or until you see it being attended by an adult bird. If you see it being attended by an adult bird, LEAVE IT ALONE. IT IS NOT ORPHANED.

If you DO NOT see it attended by an adult bird w/in the hour, then

Can you find the nest & reach it?

YES

NO

Place the bird back in its nest.

Call a licensed wildlife rehabilitator. Consult the guidelines for safe handling of wildlife and follow them.

Illustrated by Antonio Roman-Alcala

HOW ANIMALS OCCUPY THEM-
SELVES.

Ever human being, man, woman and child, suffers occasionally from what they call "the blues." No matter how happy the temperament as a general thing, a time comes in the life of each person when the "blue devils" get the upper hand. What the French call "ennui" afflicts many of us occasionally, and many fine low spirits, boredom, melancholy, the chief bane of existence. But the lower animals do not suffer at all in this way. Birds and beasts, as well as the insects, manage to pass through life without succumbing to what almost crazes their more itelligent superiors. Loafing, or chronic idleness, as everyone knows, is the most fruitful source of misery, and lack of occupation is the sure breeder of despondency. The other animals, however, do not loaf. Loafing is an art which few living creatures understand. Lizards, crocodiles and other saurians seem to have mastered the philosophy of being industriously idle without ill results, but there is nothing to indicate that they suffer from "the blues." The art appears to consist in the fact that animals have acquired the knack of making much ado about nothing; they have learned to be very busy without doing anything. But obviously this accomplishment differs from what among men is called loafing. It is one which animals have brought to perfection, and by knowing how to eliminate the horrors of "ennui" they obtain from life the maximum of happiness. Some human beings, chiefly women, also succeed in mastering this art of pottering and time consuming, though nothing greatly is being done. Old men, too, sometimes exhibit the same trait. Washington Irving remarks that one of the compensations of old age is that it allows one to be idle with impunity and without reproach. The aged rise early and are eternally "pottering around" in an aimless way, busily occupied in doing nothing of consequence.

Watch a wasp overhead busily exploring the holes in the trunk of a tree. Why he does this he probably does not know; he has not time to stop and think. He is quite content to explore away as though his life depended upon it, and will inspect over and over again every portion of the same hole. All this labor is useless in a sense, but it apparently constitutes the wasp's taste, without which it would in all probably die of what men call low spirits. Many other creatures are equally expert at frittering away time. They spend much of their time in futile dalliance. Note a canary in a cage. He hops backward and forward between two perches as though he was paid by the distance for doing so. Look at the butterfly. He leads an aimless existence, but manages to keep busy all the time and embodies the very perfection of the joy of living. A bee visits more flowers than a butterfly, but his trips are for business, while the other seems to have no other object than to keep moving. The farmer and the farmer's wife suffer more or less from worry. They have their cares, their troubles, their anxieties of a thousand kinds which often eventuate in horrible and long-continued fits of the blues." Not so the animals around them. There is no despondency in chanticleer's crow or to biddy's lively cackle. The cows and horses are always cheerful, always in good spirits and have no trouble "killing time." Unless sick or injured in some way, they are never despondent. Perhaps man's very superiorty causes much of his misery. Perhaps he would be happier if he were not endowed with reason, if he did not think. The much vaunted happiness of children is evidently the bliss of ignorance. They have not yet reached the stage of worry which comes without thinking. The lower animals never reach this stage, and their lives are free from care, from regret for the past or worry for the future. "There is nothing good or bad but thinking makes it so," remarks the melancholy Hamlet. We think ourselves into most of our unhappiness while the animals, not possessing this faculty, escape its most fruitful source of misery. The philosophy of it all would seem to be that man either knows too little or he knows too much. Do not the lower animals prove the truth of the old maxim that "ignorance is bliss?" If so, is it not folly to be wise, and is it not questionable if civilization, after all brings us any nearer to the much desired but ever elusive entity called happiness?
—American Farmer.

Guidelines For Safe Handling Of Wildlife

Megan Prelinger

1

Human safety is always the most important consideration when handling wildlife.

2

Do not offer food or water to injured or orphaned animals.

3

Do not attempt to handle:
- large birds of prey
- rabies vector species
- predator mammals
- snakes

without the assistance or supervision of an experienced wildlife care professional.

4

Have ready a box or pet carrier that is:
- dark (draped with sheet or towel)
- well-ventilated
- neither too warm nor too cool
- safe from domestic animals and curious children

5

Have on hand spare sheets, towels, and washcloths, and leather gloves for handling all animals other than small birds.

6

Gently place the injured animal in a dark, well-ventilated box or pet carrier.

7

Wildness can be disrupted by human contact. Wildness needs to be protected. At all times, minimize human contact with wild animals. This includes visual contact. Keep carriers or containers draped with loose sheeting to protect the animals from the stress of seeing people.

Do not talk to wild animals, or handle them any more than the situation demands.

8

Transfer your injured or orphaned wild animal into the care of a wildlife professional as soon as possible.

9

Be open to simplicity. If a small bird flies in to your house, you can just shoo it out again. Or pick it up and set it out gently. Brief, gentle handling will not harm it.

The Environmental
Imagination in Agriculture

Rachel Nicole Weaver

Environmental philosophy searches for models to integrate theories into practice, and endeavors to develop realistic and applicable solutions to problems in various fields. Working with the soil on a farm requires reason and physical devotion, but there are unconscious motives that affect our perceptions of food and self. Agriculture offers a means to broaden imagination and physically interact with the environment. Our understanding of food is tied up in the social and spiritual dimensions of both food and farming (Kaplan 2012, 1-21).

Encouraging the environmental imagination to interact with agriculture allows for a more nuanced model of environmental philosophy. Agriculture and food systems impact our daily lives, form part of our identities, and establish our sense of place. The environmental imagination can both enhance our understanding of the physical world and evoke unconscious engagement. Food systems augment our perceptions

of dynamic structures and interdependencies of production and processing.

A society's moral imagination shapes the food ethics and politics that become embedded in its agricultural practices. Stories, language, and narratives shape our cultural traditions—the method by which humans relate with the planet. When we interact with these methods, we can attempt to understand other histories and places in a cultural context. Agriculture presents a social domain where we can develop an ecological imagination while interacting in harmony with environments.

This imagination can both frame and mediate human and natural histories. The ability to organize and conceptualize environmental interactions requires an ecological imagination to guide our perception (Buell 1995). This imagination can be difficult to articulate however, since it lies outside our standard methods of cultural expression.

We depend on figurative language or personification to provide voice for the environment, but of course places, non-human organisms, and natural processes do not express themselves using human language. In order to become more integrated with the natural flux of the world then, we require a sensory element. The agricultural imagination can provide this element.

Using this imagination we can gain a deeper recognition of traditions, meanings, metaphorical language, and encounters with the environment. This promotes sustainable exchange between ecosystem management techniques, natural resource policies, and all living residents of a place. People, ecology, and health flourish together in an environment where participatory experiences of a place arouse the environmental imagination (Hale et al. 2011).

Cardo's Farm Project, a small urban farm located in Denton, Texas, practices alternative agricultural as performance. Young farmers work together to help a local farm and learn about sustainable agriculture. Individuals interact on the farm in order to practice personal ethics, meditate, participate, and connect with community members. Sustainable farming in Texas also illustrates the difficulties of alternative agricultural practices. Farmers must rely on ground water for the majority of their irrigation. Agricultural legislation tends to favor industrial farming models.

Organic farmers often need to find a niche market in order to have a profitable farm.

Engaging with the interplay between industrial energy and farm ecosystems employs the skills developed by the agricultural imagination. Using our cognition we can explore traditions of energy and agriculture, and use language to enhance our understandings of the relationship between food and health.

Bibliography

Buell, Lawrence. 1995. *The Environmental Imagination: Thoreau, Nature Writing, and the Formation of American Culture*. Cambridge: Belknap Press.

Hale, James, et al. 2011. "Connecting food environments through the relational nature of aesthetics: Gaining insight through the community gardening experience." *Social Science and Medicine* 72: 1853–1863. doi: 10.1016/j.socscimed.2011.03.044.

Kaplan, David M., ed. 2012. *The Philosophy of Food*. Berkeley: University of California Press.

Sanford, A. Whitney. 2012. "The Ecological Imagination," in *Growing Stories From India: Religion and the Fate of Agriculture*, 12–27. Kentucky: University Press of Kentucky.

What business can ignore the farmer's strength?

Willa Mamet *The Reader*

HERE'S HOW FARM BILL WAS PASSED BY HOUSE

By Bob Handschin
F. U. Legislative Correspondent

WASHINGTON—Here's a quick look at the 1944 Farm Bill as Senate Agriculture Appropriations committee prepares to start hearings May 10; Vital organs are missing after Farm Bureau raid: No fund for Farm Security production loans and management help; no parity payments for 1943 crops; one-third cut in benefit payments on basic and war crops; no new wheat or corn crop insurance to be written; prohibits special incentive payments already promised on war crops: War Board and AAA committee funds cut in half; no more non-recourse full insurance loans to be made; no funds for tenant purchase FSA loans: REA administrative funds seriously cut. Total funds for Farm War Budget cut by 50 percent.

Vital organs added: Maximum benefit payment cut from $10,000 to $500.

Farm Bureau dehorned on: Handing FSA program and funds over to land banks and county agents; handing AAA funds for information and campaign promotion over to county agents; eliminating all crop payments; limiting employment AAA county committeemen; taking school lunch and penny milk programs away from kids; barring any loans from land banks, PCA's, REA, banks for co-ops, commodity credit, or any other government agency unless farmer first turned down by all private lenders in his area. Senate hearings begin May 10, Farmers Union delegation preparing to appear. Under strong leadership Chairman Richard B. Russell of Georgia committee will be fighting for farmer, not Farm Bureau-crats. Our fight is to go beyond FSA budget of this year, get funds and authority for incentive payments, extend crop insurance on all war crops. Give farmer committees more responsibility. Allow non-recourse loans; get parity payment and full crop benefits; continue tenant purchase loans.

TRIUMPHS OF CAPITALISM

The amount of land you could buy with 1.3 trillion dollars ($4100/acre), the amount of the US Deficit, and, coincidentally, the US Hispanic buying power.

FEDERAL LAND AS A PERCENTAGE OF TOTAL STATE LAND AREA

Harvest Time

Jean WIlloughby

Why not occupy your days in the half shade of blueberry bushes or mulberry trees, talking and joking as you collect pound after pound of fresh fruit? Why not spend the warm nights boiling down summers' abundance and pouring it off into jars to be neatly stacked in the pantry?

On an icy morning in Denver where I'm spending the holiday with family, I get up early and head to the kitchen to examine their eating habits and piece together my breakfast. Next to a couple of store—bought jellies, I find one of the jars of mulberry-ginger jam that my partner and I made in the summer and sent to friends and relatives. My inner second-grader chimes in "Yessss! Sweet!" It had been a minor feat for us as self-taught food preservationists. As my toast begins to burn a little around the edges and the tea water comes to boil, this deep purple jar takes me back to climbing around in the biggest mulberry tree I've ever seen, planning our recipes (We'll make wine! We'll make pies! Juice! Jam!), and eating far too many berries. It occurs to me that part of the point of making things is making memories.

Do you recall how most of the items in your pantry ended up there? Do you have fond recollections about stuff you've purchased from the grocery store?

It would be too easy to dismiss berry-picking as child's play. It is child's play—unapologetically. It's messy and time-consuming—really it is time-harvesting. Once I get my arms and legs into the rhythm of the task, I'm lured by the exhilarating practicality of the whole effort. Sometimes the dirtier your hands get in making something, the more meaningful the final product becomes. It turns out money is not dirty enough to transform berries into memories.

Berry-picking: the gateway harvest.

Harvesting and preserving berries is fun, but when you dabble in garlic, or ginger, or sweet potatoes, or 100-foot-long rows of butternut squash, you're liable to get yourself hooked on growing your own food. Each crop makes unique demands on the grower throughout its season, and harvest time is no exception. Then comes the picking, the packing, the chopping, drying, pickling, smoking, salting, boiling, bottling, fermenting, freezing, and canning. Each method accentuates different flavors, leaves an aftertaste of sun and sweat, a memory of time well-spent.

It's very simple: picking, processing, storing. And it's the basis for something more complex: sharing.

As we are all aware, most Americans' lives no longer revolve around planting and harvesting, but agrarian ways of life have much to teach us. Many of us are emerging from years spent in ignorance of the plenitude that surrounds us. We're learning skills that our grandparents would have taken for granted. We find an unrivaled thrill in our rediscovery of the old mulberry trees, so many of which stand unpicked in our towns and cities. By preserving and sharing the harvest, we can revive, in the hollows of winter, a teeming summer.

Willa Mamet *Castiron*

Selected Principles of the Agrarian Trust

EQUITY FOR FARMERS

Farmers must be able to build equity in the land, and their business on it. Where ownership is not an option lease-terms must allow the farmer to invest in infrastructure, soil health and the long term interests of the property. These investments can be translated into cash value, which the farmer can retain at the end of lease term.

DIGNITY

Farm practices and economic conditions must support the dignity of labor and provide a living wage, and decent housing for farm managers, farm workers, their spouses and children. All parties have right to dignity and respect, despite differences in economic power: feudal relations are unacceptable.

AGRARIANTRUST.ORG/PRINCIPLES

Josef Beery

A Beginner's
Agrarian Reading List

Alexis Zimba Kirby

**TREATISE, ESSAYS, AND
GENERAL NON-FICTION MUSINGS**

CLASSICAL

The Works and Days, Hesiod (7th or 8th Century BC)

The Georgics, Virgil (1st Century BC)

Second Treatise of Government, John Locke (1690)

Notes on Virginia, Thomas Jefferson (1787)

Nature, Ralph Waldo Emerson (1836)

Walden: Life in the Woods, Henry David Thoreau (1854)

CONTEMPORARY

I'll Take My Stand: The South and the Agrarian Tradition,
Southern Agrarians (1930)

Look to the Land, Lord Northbourne (1940)

An Agricultural Testament, Sir Albert Howard (1943)

A Sand County Almanac, Aldo Leopold (1949)

*The Foxfire Book: Hog Dressing, Log Cabin Building,
Mountain Crafts and Foods, Planting by the Signs, Snake
Lore, Hunting Tales, Faith Healing, Moonshining, and
Other Affairs of Plain Living,* Eliot Wigginton, Editor
(1972)

The Unsettling of America: Culture & Agriculture,
Wendell Berry (1977)

Becoming Native to this Place, Wes Jackson (1996)

**NOVELS, STORIES, AND
FICTIONAL CREATIONS**

CLASSICAL

Castle Rackrent, Maria Edgeworth (1800)

Middlemarch: A Study of Provincial Life,
George Eliot (1874)

La Terre (The Earth), Émile Zola (1887)

Tess D'Urbervilles, Thomas Hardy (1891)

O Pioneers!, Willa Cather (1913)

CONTEMPORARY

The Good Earth, Pearl S. Buck (1931)

Light in August, William Faulkner (1932)

The Grapes of Wrath, John Steinbeck (1939)

Prodigal Summer, Barbara Kingsolver (2000)

That Distant Land, Wendell Berry (2004)

A NOTE ABOUT CLASSICAL VERSUS CONTEMPORARY:
This distinction is highly debatable and based on the author's
personal distinction between "traditional" agriculture, tbefore the
Industrial and Agricultural Revolutions of the 1930s when draft
animals were the main source of farm power, and "industrial"
agriculture, after the normalization and widespread use of gasoline
powered machinery on farms beginning after the 1930s and leading
up the Big Ag farming of today.

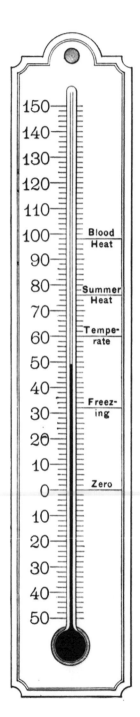

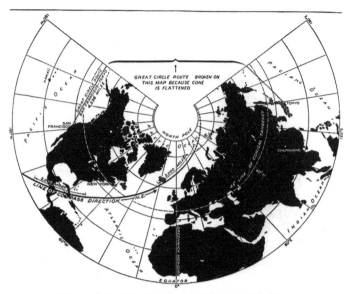

Full spread yoke of Simple Conic Projection. Note spiral of line of constant oblique direction °

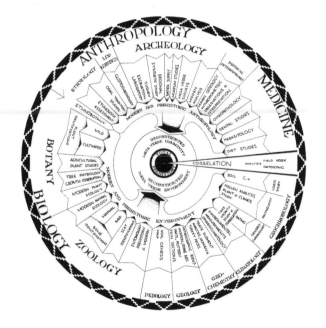

"Conservation is getting nowhere because it is incompatible with our Abrahamic concept of land. We abuse land because we regard it as a commodity belonging to us. When we see land as a community to which we belong, we may begin to use it with love and respect."
—Luna Leopold, [intro to]
A Sand County Almanac, 1966

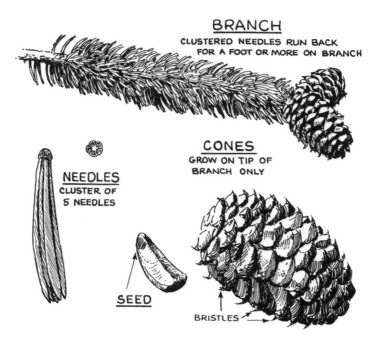

BRANCH
CLUSTERED NEEDLES RUN BACK FOR A FOOT OR MORE ON BRANCH

NEEDLES
CLUSTER OF 5 NEEDLES

CONES
GROW ON TIP OF BRANCH ONLY

SEED

BRISTLES

EDITOR'S NOTE Bristlecone pines, found on the arid heights of the Great Basin's desert mountains, can grow to be over 4000 years old. It helps put into perspective some of our own human preoccupations when we consider, for example, that all the food systems we've built, wrecked, and re-built in much of human history have happened within the lifetime of one tree.

July

Start-Up &
Steadystate

Chapter 7

Cooperative Movements and the Grange

Jen Griffith

Drive down a rural road or through a small farm town and you have a good chance of bumping into a Grange Hall. The word "Grange" evokes an era of vibrant and organized agrarian communities, but few know how or why these buildings spread across rural America or what activities they housed. The history of the Granges and the plight of the farmers who built them are unique to their era, but contain themes familiar to all farmers.

The Grange, or the Order of Patrons of Husbandry, was founded in 1867 to support farmers and rural livelihoods. At the time of its inception, America was recovering from the Civil War, railroads were spreading, farms were industrializing, the West was just barely settled and women's role in society was changing. In an attempt to improve the lives of farmers and uplift the profession within society, the Grange fought the railroad with anti-monopoly laws, advocated for rural mail delivery, supported women's suffrage, included women as full voting members, united southern and northern farmers, and helped shape the recently formed Land Grant institutions to serve farmers' needs.

On top of all of these other accomplishments, the Grange funneled time and resources into developing cooperative business ventures among its members. Farmers were fed up with the monopolies formed by railroads and warehouses, and members were attracted to the Grange as a vehicle to rise up against these businesses and regain some control in the market.

TYPES OF ENTERPRISES

Collective Purchasing and Selling

Local Granges created agencies through which agricultural products, equipment, and household goods could be bought collectively at lower prices. Sometimes the agencies would band together at the county or state level to increase their clout. Farmers across the country collectively ordered farm implements, seed, sewing machines, parlor organs, wagons, feed grinders, shellers, hayforks, clothes, groceries, etc. Farmers also pooled common crops and sold them collectively, getting better deals by working in larger volume.

Cooperative purchasing proved to be both a success and a failure. Many of the projects were ruined by the combination of overzealous and under-experienced leadership. Instead of growing slowly and steadily, in many cases businesses grew faster than the agencies could properly manage. A few cooperatives persisted for many decades—notably Agway, which went bankrupt only in 1993. Because of competition from cooperative purchasing, the period saw an overall reduction in prices benefitting all farmers and lowering the costs of necessities for everyone.

Manufacturing

When Grangers tried to purchase larger equipment collectively, many manufacturers would not sign contracts with the organization. They had already

established relationships with regional stores who had exclusive rights to sell their products. As a solution, some Granges decided to build their own factories and produce their own machinery to break up these "Harvester Rings" and "Plow Rings."

Grange manufacturing businesses across the nation proved to be one of the biggest failures in the cooperative movement. Businesses were poorly managed, and factories built defective machinery or shipped product too late in the season for farmers to use. Larger businesses sued the Grange companies for patent infringement. Within a few years, many of these manufacturers closed.

Smaller-Scale Infrastructure Development

Grangers found much greater success in developing smaller, more locally-minded infrastructure that filled a need within their particular farm community. Granges collectively built grain elevators, grist mills, cheese and butter factories, linseed oil factories, starch factories, pork-packing establishments, hemp factories, cotton mills, and creameries.

These operations were major successes for the Grange. The businesses were smaller in scale, more manageable and provided a local service within the skill set of the farmers. The endeavors had greater longevity and did not financially drain the Granges.

Banks/Lenders

In the 1870s farmers felt that the banks discriminated against them, charging high prices and not offering the loans needed to run their businesses. California was the first state to initiate a solution by creating the Grangers

Bank of California. Only Grangers could hold stock or serve as directors of the bank.

The successful bank was instrumental in helping farmers through the financial turbulence of the time. When the wheat market collapsed in the mid-1870s the bank loaned money to farmers so they could hold onto their wheat until prices rose again.

Insurance

Granges across the country organized insurance agencies, usually at a very small and local level though a few larger, statewide programs were created. Fire insurance and life insurance were the two most common types created.

The Grange experienced both success and failure in organizing cooperatively. Reaching too far too quickly was an early cause of the failures. But there were also many successes. Grange stores and buying clubs put pressure on other retailers to lower prices, benefiting not just the Patrons of Husbandry but everyone. For some ventures, the need was clear and the farmers had the knowledge to be successful.

Living in an era of a consolidated food system and governmental policies that favor large corporations, there are many parallels between the plight of today's farmers and the Grangers. Their struggles in the 1800s can remind us of the possibilities of cooperative movements.

CO-OPERATIVE RESERVE
UNITED TO ASSIST, NOT COM-
BINED TO INJURE.

CHARACTER RESERVE
SELF-GOVERNMENT RELIGION EDUCATION HOME

HEALTH RESERVE
WELL BODIES, SOUND MINDS
FIRM WILL, HAPPY SOULS

GOLD RESERVE
$1,850,000,000

LAND RESERVE
CROPS WORTH TEN BILLIONS
3,623,000 SQUARE MILES

Fivefold Reserves of the American People

*T*HE future of the United States as a nation depends upon the use which the American people make of their fivefold reserves:

1. Character Reserve—mind, will, soul, as expressed in home, education, religion, self-government and industry.

2. Health Reserve—physical, mental, spiritual.

3. Co-operative Reserve—the cohesive power of individual effort when wisely associated for the common welfare.

4. Gold Reserve—making every dollar in exchanges as good as gold.

5. Land Reserve—mobilized as the basis of all wealth.

While air, land and water support all life, the real power of America lies in the extent to which each of these five reserves—Character, Health, Co-operation, Gold and Land—are harmoniously inter-related, inter-conserved, collectively utilized. Each of these five forms of national reserve must be made strong, none of them can stand alone—they must all develop in unison and operate in harmony in order to produce the largest and best results in individual and corporation, state and nation.

Without character and health, neither co-operation, money nor land will avail much. But associate the three latter with the two former, and the combination is rich in endless possibilities of human endeavor, achievement and happiness. Granting this premise, the Myrick method aims to outline a feasible scheme for associating land, money and human effort for the largest efficiency.

Blame It On The Blizzard Of '78

CR Lawn

The thick grey blanket of clouds moved in on Feb. 6, 1978. Then came the snow, driving in almost completely horizontally from the east, in thick flakes that would accumulate to 18 inches. I was living in an uninsulated, self-built 20x14' cabin on a discontinued town road in South Canaan, Maine, with no electricity or plumbing, and two wood stoves. In the midst of the tempest my stove pipes, secured by remarkably strong discarded guitar strings, blew down. My lone resort was to brave the blizzard and cross the road to the only other cabin on the 52-acre homestead.

It was my third winter in this relative isolation. It would be my last. At age 31, I had had enough; before the next winter I would make the Maine Federation of Co-operatives an offer they could not refuse. For the princely sum of $75 per month, I would come live in the office in the back of their food warehouse for the months of December, January and February to work on special projects.

Fedco Seeds (Fedco is short for "Federation of Co-operatives") became one of those special projects. It was modeled after Northern Lights Provisions, which pre-ordered seasonal staples, and had been a part of the burgeoning homesteader-driven Maine co-operative movement of the '70s. Garden seeds, for most folks a one-time yearly purchase, were perfectly suited to such a model. The existing Maine Federation of Co-operatives, with its scores of buying clubs, offered a ready-built market. The seven-person collective running their food warehouse were potential consultants with the brains and experience to help me.

I went around to each, picking their brains and learning. I quickly realized that their weekly collective meetings were deadly. Every single detail, every single decision had to be by consensus, with no delegation, no trust. Instead of being inspired, people came out enervated.

Moreover, the Federation, at its first meeting, had made a dreadful decision that would ultimately doom it. Based on some fuzzy thinking and dogmatic ideology about not allowing the capitalist system to exploit co-op labor, they would sell only to other co-ops, not to individuals or for-profit

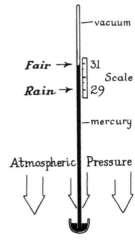

BAROMETER

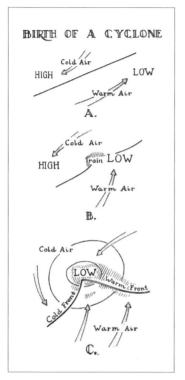

BIRTH OF A CYCLONE

HIGH — Cold Air — LOW

Warm Air

A.

HIGH — Cold Air — rain LOW

Warm Air

B.

Cold Air

LOW — Warm Front

Cold Front

Warm Air

C.

but instead by brainstorms involving the few people—usually two to four—who had the most at stake in any given project. I called such decision-making by small get-together the consultative strategy.

We accepted individual orders right from the start too, though with a $25 minimum order at first. In time we got rid of the minimum and accepted orders of even a single packet. In place of a policy of exclusivity, we had formulated one of inclusion. That's why we sell conventional as well as organically produced seeds and try to appeal to every sort of gardener and farmer: from postage-stamp-sized urban gardens to vast homesteads, from CSA operations to wholesale market farmers.

Already, I had figured out the seed business. You could buy a small 5 gram packet of beet seed for 45 cents. You could buy a whole pound of the same seed for $3.40. Why not break down that huge packet into small units using co-op labor and resell them? Our first year we found

we could sell a 5 gram packet for a dime and still make a profit.

The idea was too good to remain long within the confines of Maine. In the next year, Fedco Seeds would spread like wildfire through Massachusetts and then across the remainder of New England, tripling our volume and then doubling it again the following season.

I never dreamed that I had found my lifetime livelihood. Nor did I ever imagine that one day we would sell nearly three quarters of a million seed packets per year, run five separate businesses, generate nearly $5 million sales annually, and employ 30 full-time employees, with sometimes as many as 70 on at peak season. I could not have predicted how internet ordering would hasten our growth, turn us into a nationwide cooperative, and revolutionize order processing. And now we're shipping directly to Canada, too.

That storm blew in a lot more than snow!

businesses. Fedco Seeds would not work that way, I vowed. I developed a meeting-averse culture. Creative ideas, I believed, usually came not from large or even small meetings,

"Let your countenance be pleasant, but in serious matters somewhat grave."

"Let your discourse with men of business be short and comprehensive."

—George Washington, *Rules of Civility*

Time In July

Douglass DeCandia

A pause between the
out and in
breath
of summer.
This heat
brings with it
the quiet
that has settled in the fields,
distilling passions and lust into Love;
the becoming of fruit from flowers.
This time of Light
in a moment of silence.
Soon to welcome the harvest.

Serviceberry

Mycology You Can Get Your Hands Around:
Six Lessons Of Fungi

Wesley Price

Healthy soil is one of the most complex systems imaginable. Cyclical growth and decay gives rise to a biological symbiosis of interconnection which we are just beginning to understand, and which we depend upon for not only survival, but for happiness. Healthy soil and happiness? Yes, exactly.

Complexity aside, there are innumerable ways to co-create a healthy soil. The focus of this article is just that: an exploration of how we may leverage our growing understanding of fungi for long-term planetary sustainability. How? Good question. The idea that we can, through the intelligent utilization of fungi, become stewards of the soil is at its heart a revolutionary ideology—an ideology which humbly recognizes our inseparable interconnection with nature. The field of mycology (the study of fungi) offers us a solid reference point for this task of building and maintaining a healthy soil.

When seeking to integrate farming and fungal ecology, one can recognize the importance of biomimicry in establishing best practices for farming. The Biomimicry Institute explains that "The core idea is that nature, imaginative by necessity, has already solved many of the problems we are grappling with. Animals, plants, and microbes are the consummate engineers. They have found what works, what is appropriate, and most important, what lasts here on Earth. This is the real news of biomimicry: After 3.8 billion years of research and development, failures are fossils, and what surrounds us is the secret to survival."[1]

What then can we learn from the Kingdom of Fungi?

At first study, fungi may appear primitive, but do not be fooled. The fungi have carved out a niche which effectively underpins the whole of life on earth. Mycorrhizal fungi, by definition, have evolved to associate with the roots of plants. This mycorrhizal association is understood to confer drought resistance, increase the transport of water and essential nutrients and help plants fight off pathogenic infection.[2] In order to get your hands and head around this living art form, let's consider Six Lessons of Fungi, which I have adapted from the "Seven Lessons of Schoolteaching," as written by educator John Gatto.[3]

1   The Biomimicry Institute, "What is Biomimicry?" http://www. biomimicryinstitute.org/about-us/what-is-biomimicry.html

2   "Mycorrhizal Associations: The Web Resource," http://www.mycorrhizas.info/roles.html

3   John Gatto, *Dumbing Us Down*

1. The first lesson fungi teach is patterns.

Everything fungi teach can be understood as a pattern of nature. Even in the harshest ecologies a close examination of fungi turns up a coherence, which is full of internal repeating patterns, similar to a fractal. This coherent paradigm of patterned thought is thrust upon us by fungi, each working in relationship within the soil-food web. Fungi teach us how to understand patterns.

2. The second lesson fungi teach is fluidity.

Fungi interact within their ecological niches with a skilled enzymatic precision that enables digestion of nearly any imaginable food source. They act like nutrient rivers within the soil matrix.

3. The third lesson fungi teach is awareness.

Fungi must recognize and adapt to changes in their environment. They must turn on and off like a light switch, when the conditions for growth are right. Fungi accept each undertaking with a keen awareness of their surroundings.

4. The fourth lesson fungi teach is sovereignty.

Once they have established themselves they are able to work to colonize and control specific nutrient resources necessary for their survival. This may mean that the host plant is improved, as in many mycorrhizal relationships; or the host plant is destroyed, as in parasitic relationships; or that the host plant is slowly used as a food source, as in saprobic fungi.

5. The fifth lesson fungi teach is the interrelatedness of everything.

Fungi teach interconnections. It is the most important lesson: that we are completely embedded in the natural world and as conscious, sentient beings, we are alive within it.

6. The sixth lesson fungi teach is that one can't hide.

Fungi are always watching us, waiting for the opportunity to recycle us and continue the nutrient cycle.

GILLS

VEIL MAY OR
MAY NOT BE PRESENT

(NOTE ABSENCE
OF BASAL CUP)

CAP

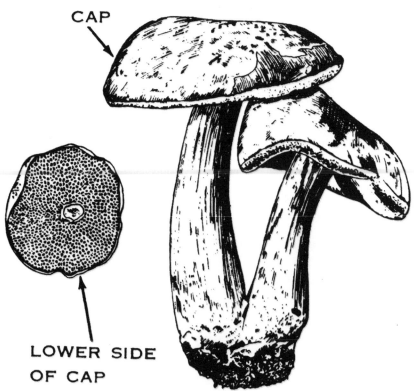

LOWER SIDE
OF CAP

Try To See No Evil

Rebecca Gayle Howell

From a distance the brushfire looked like veins crossing,
a flame's thin lines, like electric wires, like Christmas.
Cigarette ash, a knoll of grass; these days the air is dry
enough to spark. Sometimes I dream the ground approaches,
soil of copper snakes, tongues pronged as divining rods
whipping the air. Last night two diamond backs, a cross of tails
bent to the motion of a clock. Tey moved down my neck.
I woke cold and sweating. I woke like I didn't have a home.
Tey say the low moon is coming. Maybe so. God knows
why else the street dogs would be making so much noise.
I wish it meant something. I wish a colored moon could pull
so strong dirt would gush a well of water. I'd get my silver bucket.
I'd open my mouth. Te fire will be all right. It's a game;
one guy sets it from boredom and from boredom the other
puts it out.

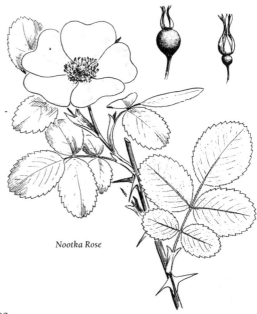

Nootka Rose

Some Useful Suggestions for Beginning Homesteaders

Robert Winfield Freeman

Grow sweet potatoes along a stone wall, or put large granite stones along your sweet potato rows. They heat the soil, and this increases yields of sweet potatoes, a tropical crop.

Drive a T-post into the ground, and bolt a second post to it upside-down so it's about 11 feet high. Do the same thing about 7 feet away, and another one another 7 feet away. Secure a length of cattle panel to the upper end of the fence posts. Now you have a trellis fence that is 11 feet high. Hang string off these to tie up tomato plants, and train those tomato plants 11 feet high. Use TreeGators around the tomato plants to keep them consistently watered and prevent blossom end rot.

Kale, cabbage, beets, spinach, broccoli, and cauliflower can all be started in pots, close together. They are very hardy to transplant.

Gather your neighbors' discarded leaves, pile them up and put some horse manure on top of them. Let it rot down: instant dirt. No need to turn, in my opinion, the worms turn the compost. Also, use leaves to mulch down weeds. If necessary, you can mine a hardwood forest for leaf matter.

If you can get the kind of pallets that don't have any gaps in between, nail them together into a four-sided box without bottom or sides. Fill it with organic matter with cordwood at the bottom, because that cordwood will sop up water and water the plants for a few years. Now your Hugelkultur box garden is 3 feet off the ground, and you don't have to bend over to weed it. It's a good way to introduce gardening to children or to people who can't or don't want to bend over all the time. It also keeps a border between garden and not-garden, which is a big problem for traditional row gardeners when the weeds get out of control.

Burn campfires on your garden during the winter. Burn dried plant matter on your garden (of course with a hose at the ready and permission from the fire marshal). Ashes and partially burnt wood are very good for the soil and the fire might kill some weed seeds.

EDITOR'S NOTE Acorns, hickory nuts, honey locusts, and other 'tree crops' drop what's called 'mast,' a powerful reserve for fattening hogs and goats. The value and ecology of the hedgerow has been much described by Oliver Rackham, an English author not distributed in the States. Re-diversifying our hedgerows, fencerows, and buffer areas will take at least our own lifetime. See Connor Stedman's remarkable guidebook on Agroforestry, published by the Greenhorns and available online for free at www.thegreenhorns.net.

"HE THAT BLOWETH NOT HIS OWN BUGLE, THE SAME SHALL NOT BE BLOWN."

LARNED, PAWNEE COUNTY, KANSAS, FRIDAY, DECEMBER, 7, 1888.

FROGS NOT PROFITABLE.

Five Thousand Dollars a Year Cannot Be Made Raising Them.

In the Boston *Journal* some weeks ago appeared a quotation from another paper regarding a "frog farm," in which the extravagant statement was made that an income of $5,000 could be obtained from raising these inhabitants of the marsh for sale. Fred Mather, superintendent of the New York Fish Commission and fishery editor of *Forest and Stream*, demolishes this assertion by an interesting article in the last issue of the *Forest and Stream*. Therein he writes:

For the past four years this question of frog culture has given me some trouble in answering private letters from would-be frog farmers, and in a legacy left by the late Seth Green, who wrote an article on it in the report of the New York Fishery Commission for 1873. This was widely copied, and each year some imaginative reporter gives an account of a mythical frog farm which has never existed. In 1875 this farm was located near Smithtown, Long Island, and I went there and found that there was no frog farm in that vicinity, nor was there any man living near the same place bearing the name given. The next year the apocryphal farm was located near Philadelphia, and I had the same experience. Since that time I have only wasted postage in the pursuit of the fabled industry.

Mr. Green gave the results of two years' experience in which he gathered frog spawn and hatched it, but lost his polywogs and abandoned frog culture, but he encouraged others to try it by saying the "difficulties can be overcome by patience and perseverance." Yet he further said:

"When they become frogs they live on all kinds of insects, and the only thing I can see to make success sure is to procure insects in large quantities, enough to support a great number of frogs." And here is where the impracticability of raising frogs lies.

It is no trouble to gather large quantities of frog spawn and hatch millions of tadpoles, nor to feed the latter on meat and vegetation, for they eat both. The enemies of the tadpole, or polywog, are numerous, and large frogs will eat small ones. As an instance of the latter fact, I once took a dozen large American bullfrogs to Professor Moore, of the Derby Museum at Liverpool, and they quickly swallowed the small European frogs that were in the tank where he placed them. Even if it were impossible to feed the frogs on meat, which it is not, the batrachian is of slow growth, and the balance would be on the wrong side of the ledger when they were sent to market.

I will travel to see the frog farmer "living on an income of $5,000," all from his frogs. In my long experience as a pisciculturist the frog has been under constant notice, and I have tried to feed them on meat and mussels without success. My belief is that frog culture is a delusion, and that such a thing as a frog pond does not exist, and unless some genius arises who can find a way to feed his frogs on beetles, flies and other insects, which, with some snails, constitute their natural food, there will never be a frog farm. The supply of frogs to New York markets comes mainly from Canada and places of sparse population, where they are not extensively eaten. Forty years ago Americans did not eat frogs, and every marsh in the country contained large ones. Now they are almost extinct near cities in the State of New York, and do not breed and grow rapidly enough to pay to catch them. I think it probable that the batrachians, like the reptiles, are of long life and slow growth, and that a man would get very tired waiting to see a crop mature. Perhaps he might wait ten years to get large ones, but never having raised a frog to maturity I can not say. The belief in their slow growth is based on the fact that many polywogs do not get their legs the season they are hatched, but pass the winter in the tadpole state; this points to slow maturity.

10

4

1

About the Agricultural
Adjustment Program

Cambridge, Mass.
March 18, 1942

Mr. E. K. Dean,
Kansas Union Farmer
Salina, Kansas
Dear Mr. Dean:

I will appreciate your running the enclosed letter in your column in some space where it will be noticed. I want to find out what farm people think the farm production program should be like next year. I am sure they can tell me much better than I can find out from my friends in the government service.

Very truly yours,
John D. Black,
Professor of Economics,
Harvard University.

Dear Sirs:

What should be done with the Agricultural Adjustment program in 1943?

This is a highly important question upon which farm people everywhere should be doing some hard thinking—and doing it soon, for the government fellows are already at work on it. Congress is even now beginning to pass legislation that affects next year's Adjustment program. The administrators in federal, regional and state offices are already pushing their pencils around making plans for next year's crop.

The goals that have been set for this year's agricultural output average 6 percent more than the 1941 output. It won't be easy to reach these. But 1943 is going to see the need for amounts so much larger than this 6 percent that one's heart almost stands still to think of it. Can our agriculture do it? How?

We must not let the government folks do all the figuring on this. There are more good ideas on a question of this kind outside the government than in it.

In fact, it is not even safe to let the government folks do all the figuring about it. They're just human beings and can not keep from thinking too much in terms of just more of what they are now doing. It's their job.

For some twenty years now, nearly all of the time as an outsider, I have been trying to understand agriculture's problems and help work out ways of meeting them. Some of you know that I had a bit to do with developing the present Adjustment program. I am much concerned whether it should be continued next year, and while the war lasts, just as it is now? or whether it should be changed? And if so, how?

What I would like is that some of you who read this letter would write me rather soon your ideas as to how the present AAA and production goal program is working out in your area? And whether continuing it as it is will give us the increased output we need in 1943? Or whether some changes should be made? And what these should be made? Should any of the restrictions in the present program be removed? In fact, should there be any restrictions at all? If so, which ones should be retained?

I know about the shortages of labor in some sections already. It will be worse in 1943. What ideas have you about meeting this problem?

Is lack of credit going to hold anybody back?

I don't promise to answer fully and individually all the replies which I receive to this request. But I shall study them all carefully and make very good use of what is in them. They will be treated confidentially—none of them published, at least without your permission.

Very truly yours,
John D. Black,
Professor of Economics,
Harvard University

Try To Speak No Evil

Rebecca Gayle Howell

I smell the storm before I see it, but I do see it.
Not the rain, but its expectation, the light's
filtered wait. Faces, store signs, my neighbor's car—
Polaroids, envelopes of subtractive color, slow
focus. Change is certain except when it's not.
Anymore, I can't tell if I'm awake. How about now?
Once I saw a twister of birds, black raptors,
slide along a hidden helix right down to the white sun.
They did not burn. Here was a thing, I thought,
that did not move on.

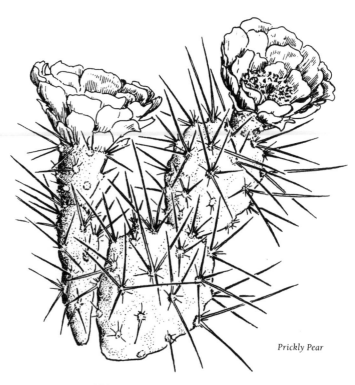

Prickly Pear

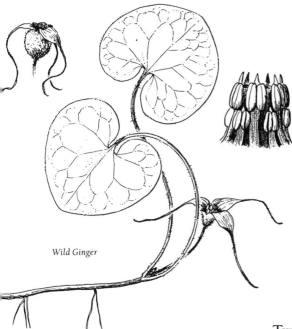

Wild Ginger

Try To Hear No Evil

Rebecca Gayle Howell

The old factory's windows were broke out, the inside
gutted. Plaster cracks on the load baring walls
snaked under each other exposing the lath beneath,
ordered and nailed. The cladding was chipped
brick and mortar, and where once the roof laid down
rose the branching of chestnut trees; trunks thrust up
through the concrete floor. I stood, small. I knew
how much I could not hold. The meeting house.
In the dream I wasn't clear what we'd come to pray for,
but that we'd come, each of our own terror nights,
to say something that could be heard. Once inside,
instead of words, I felt a hum, not a song but a groan,
like the one my mother made when she worked.

August

REAP THE BOUNTY THAT YOU SOW

Technology

Chapter 8

Birdsfoot Trefoil Makes a Comeback

Anita Deming

Birdsfoot Trefoil, (trefoil), *Lotus corniculatus*, is a plant of European origin that was intentionally introduced to the Hudson Valley as a forage crop. Like alfalfa and clover it is a leguminous plant that can fix nitrogen from the atmosphere so it needs less and cheaper fertilizer than grass forages.

Trefoil grows from 5 to 20 cm tall depending on growing site and fertility. It grows from a crown with fine wiry branches. The leaves are small in groups of three with two broad leaflets at the base of the leaf branch. The flowers are bright yellow with four to seven pea-like flowers in a cluster. When ripe the seed pods are long and brown. The seed pods give the appearance of a bird's foot as they radiate out from the flower stem.

Birdsfoot Trefoil has many advantages—it is drought resistant, tolerant of wet soils, and grows well in lower pH soils than alfalfa. It will persist for years in hay or pasture situations. Since it is drought resistant it grows in the summer heat and can provide summer pasture. It seems to do particularly well in clay soils. However trefoil will NOT tolerate ponded water or ice sheeting. An especially valuable feature of Trefoil are the tannins, which provide an anti-bloat factor that is important for ruminants (cows, sheep, goats, etc.) on pasture. Another advantage is that trefoil can reseed in favorable conditions, which neither alfalfa nor red clover will do.

To plant trefoil you need a fine and firm seedbed as the seeds are small. The soil should be a pH of at least 6.0. Try to control weeds before planting. If using a "stale seedbed" to control weeds, plant right after the last disking. Planting after corn or another row crop will also control weeds such as quack grass. Trefoil seedlings need almost as much phosphorus as alfalfa. Have your soil analyzed using the Morgan solution to determine how much phosphorus and other nutrients you need to apply.

Unless the trefoil seed is pre-inoculated, use a Rhizobacterial inoculant with a sticker such as sugar to ensure you have enough of the symbiotic bacteria that attach to the roots of legumes and fix nitrogen from the atmosphere. It is especially important to use an inoculant if you are planting on a field that has not had trefoil on it before. An alfalfa/clover inoculant won't do the trick—you need to use the inoculant specific for trefoil. Growing trefoil and other legumes is a "green" way to reduce greenhouse gasses compared to fertilizing grasses, since with grasses you need to supply external nitrogen that is more likely to leach or volatilize from your fields. The Rhizobacteria take nitrogen right out of the atmosphere and provide to the roots of the legumes directly.

Planting trefoil with an improved grass to fill in and provide additional forage is recommended. For pasture

Birdsfoot Trefoil seedpods

Both trefoil and alfalfa begin their first spring growth from stored energy in their taproots. However, alfalfa will replenish its root reserve in about 35 to 45 days throughout the growing season. Trefoil is different as it depends on photosynthesis from remaining leaves to supply the energy for summer regrowth. After about the first of September, trefoil once again will attempt to rebuild its root reserves. Controlled continuous grazing (do not overstock) that never completely removes all the trefoil leaves from the plants gives better results than complete defoliation followed by a rest period (rotational grazing). Do not graze below 4 inches and be sure to keep some leaves on the stems.

Trefoil's way of storing root reserves explains why the removal of summer growth by heavy grazing will result in stand failure. It is critical to leave some leaves on the stem all summer long. Allow trefoil to go to seed at least every 3 years to keep the stand strong. Harvesting or grazing between September 1 and the first killing frost is not recommended as this period is needed to allow root reserves to accumulate nutrients to improve winter survival and growth the following spring. Spittle bugs and leaf hoppers can be a problem on trefoil. It is also susceptible to crown and root rots.

The most important feature of trefoil is its long stand life once established. Most recently, as soil health has improved on farms, and alfalfa management has become more intensive, trefoil is used for erosion control on steep slopes near highways. Its deep rooting, longevity, lower pH requirement, ability to grow in wet and dry years, and no need for nitrogen fertilizer make it a clear asset for wild situations. In addition its spectacular yellow flowers are a joy for passersby and an excellent source of nectar for bees both domestic and wild. Additionally Trefoil will reseed if the seed pods are left to ripen. This will thicken the stand and extend its useful life.

use 5 pounds of trefoil plus 1 pound of bluegrass and 3 pounds of orchard grass or bromegrass. For hay use 5 pounds of trefoil and 5 pounds of bromegrass or timothy. As long as soil conditions are adequate for planting, the earlier you plant the better. Plant anytime from early spring until the end of July, avoiding planting when soil conditions are unusually dry. Plant ¼ to ½ inch deep (deeper in sandy soils), and firm the soil during or immediately after seeding for good seed-soil contact. We recommend using a grain drill with press wheels, or a grain drill followed by cultipacking. Trefoil is a slow grower early on, so you can use a "nurse crop" such as oats. Mowing your new seedings early to remove weed competition is critical in the first 90 days.

Trefoil regrows from the tip of the stem, so several cuttings are not expected as with alfalfa, however late cuttings are just as nutritious as earlier cuttings. A typical yield is about 2 tons to the acre. It is hard to make good dry hay from trefoil as the stems are hard to dry down, and by the time they are dry the leaves will fall off (shatter) when baled. Most of the nutrients are in the leaves, but if the stems are wet when baled the hay will mold.

EDITOR'S NOTE After watching hundreds of acres of Adirondack perennial pasture being torn up for GMO corn, I started a conversation with my local extension agent about historic crops in the area. She wrote this article in response to those questions.

8th Month AUGUST, 1957 31 Days

MOON'S PHASES —1h = C.S.T. —2h = M.S.T. —3h = P.S.T.	Eastern Time D.	H.	M.	Sun on Meridian Civil Time D.	H.	M.	S.
☽ First Quar.	2	1	55E	1	12	6	13
○ Full Moon	10	8	8M	8	12	5	32
☾ Last Quar.	18	11	16M	15	12	4	22
● New Moon	25	6	32M	22	12	2	48
☽ First Quar.	31	11	34E				

See Explanation of Calendar Pages as to times given

Days		Astronomy, Church Days, etc.	MOON'S PLACE S	Calendar for Northern States: Sun rises H.M.	Sun sets H.M.	Moon sets H.M.	Calendar for Southern States: Sun rises H.M.	Sun sets H.M.	Moon sets H.M.	Moon's Southing or Meridian Passage Eve. H.M.
1	Th	♀ sets 8:34 eve. ☌ ♆ ☽	♎	4 44	7 27	10 19	5 09	7 03	10 35	5 01
2	Fr	Gemini rises 2-4 morn. ☊	♏	4 46	7 26	10 57	5 09	7 02	11 17	5 54
3	Sa	Scorpio sets 11 eve.-1 morn.	♏	4 47	7 25	11 37	5 10	7 02	morn.	6 46
31.		Seventh Sunday after Trinity. Mark 8.		Length of Day 14h 36m—of Twilight 2h 5m						
4	Su	N. Crown sets 2-3 mo. Cl. ☌ ♄ ☽	♐	4 48	7 24	morn.	5 11	7 01	12 02	7 40
5	M	Unuk sets 1:12 morn. ◡	♐	4 49	7 22	12 24	5 12	7 00	12 51	8 31
6	Tu	Transfiguration	♐	4 50	7 21	1 15	5 12	6 59	1 43	9 22
7	W	Alpheratz merid. 3:08 morn.	♑	4 51	7 19	2 09	5 13	6 58	2 36	10 12
8	Th	Job's Coffin merid. 11:30 eve.	♑	4 52	7 18	3 07	5 14	6 57	3 31	10 59
9	Fr	Aries rises 9-11 eve.	♒	4 54	7 16	4 06	5 15	6 56	4 26	11 45
10	Sa	Gemini rises 1-3 morn.	♒	4 55	7 15	rises	5 16	6 55	rises	morn.
32.		Eighth Sunday after Trinity. Matt. 7.		Length of Day 14h 18m—of Twilight 1h 59m						
11	Su	Dog Days end	♒	4 56	7 14	7 22	5 16	6 53	7 14	12 29
12	M	Moon Apogee	♓	4 57	7 12	7 48	5 17	6 52	7 45	1 11
13	Tu	Corvus sets 7-8 eve.	♓	4 58	7 10	8 11	5 18	6 51	8 14	1 53
14	W	Andromeda merid. 2:30-4:30 mo.	♈	5 00	7 09	8 37	5 18	6 50	8 45	2 35
15	Th	Assumption B.V.M.	♈	5 01	7 07	9 04	5 19	6 49	9 17	3 17
16	Fr	Denebola sets 8:59 eve.	♈	5 02	7 06	9 35	5 20	6 48	9 52	4 01
17	Sa	Leo sets 8-10 eve. ☋	♉	5 03	7 04	10 10	5 21	6 47	10 31	4 46
33.		Ninth Sunday after Trinity. Luke 15.		Length of Day 13h 59m—of Twilight 1h 54m						
18	Su	Aries rises 10-11 eve.	♉	5 04	7 03	10 52	5 22	6 46	11 16	5 34
19	M	Arcturus sets 11:35 eve.	♊	5 06	7 01	11 39	5 22	6 44	morn.	6 25
20	Tu	Cetus merid. 2-5 morn. ◠	♊	5 07	6 59	morn.	5 23	6 43	12 06	7 19
21	W	Jupiter sets 8:13 eve.	♋	5 08	6 58	12 37	5 24	6 42	1 04	8 16
22	Th	♎ sets 9-11 eve. Close ☌ ♀ ♃	♋	5 09	6 56	1 41	5 25	6 41	2 07	9 13
23	Fr	Algol rises 7:54 eve. ☌ ♅ ☽	♌	5 10	6 54	2 52	5 25	6 39	3 13	10 12
24	Sa	Virgo sets 8-10 eve.	♌	5 11	6 52	4 08	5 26	6 38	4 24	11 09
34.		Tenth Sunday after Trinity. Luke 19.		Length of Day 13h 38m—of Twilight 1h 50m						
25	Su	Moon Perigee	♍	5 13	6 51	sets	5 27	6 37	sets	ev. 06
26	M	♒ mer. 11 ev.-1 mo. ☌ ☿ ☽, ☌ ♂ ☽	♍	5 14	6 49	7 06	5 28	6 36	7 12	1 02
27	Tu	Lyra mer. 8-9 eve. ☌ ♀ ☽, ☌ ♃ ☽	♎	5 15	6 47	7 43	5 28	6 34	7 50	1 57
28	W	Betelgeuse rises 1:04 morn.	♎	5 16	6 46	8 18	5 29	6 33	8 31	2 51
29	Th	Auriga rises 11 ev.-1 mo. ☌ ♆ ☽	♏	5 17	6 44	8 56	5 30	6 32	9 15	3 46
30	Fr	♑ merid. 10-12 eve. ☊	♏	5 18	6 42	9 36	5 30	6 30	9 59	4 40
31	Sa	♄ sets 10:38 e. Cl. ☌ ♄ ☽, □ ♄ ☉	♏	5 20	6 40	10 22	5 31	6 29	10 48	5 34

GENERAL WEATHER PREDICTIONS FOR AUGUST, 1957.—*1st to 3rd.* Variable spell. Pleasant in eastern states, becoming overcast and blustery. Dangerous storms on the Pacific slope and in northern Rockies, spreading eastward to the Mississippi valley. *4th to 7th.* Storm period. Hurricane threat to the southeast, slowly clearing in central and western states. *8th to 11th.* Fair weather. Thunder storms in eastern areas, fair in the Pacific states with fogs on the coast. *12th to 15th.* Heat wave. Fair and hot in the west, becoming unsettled. Clear skies in the Mississippi valley and over the prairies, clearing and cooler in the upper Atlantic area, dry in southwest. *16th to 19th.* Storm period. Blustery in the Mississippi valley, squalls in the northeast and thunder storms in the Gulf states. Stormy in west coast states with violent winds. *20th to 23rd.* Pleasant spell. Mostly sunny in central and mid-western states, very windy along the west coast. *24th to 27th.* Unsettled weather. Thunder storms in the south, unsettled and blustery in the Ohio valley. *28th to 31st.* Storm period. Gales along the Pacific coast and stormy east of the Rockies, spreading to the Atlantic by the end of the week. Precipitation below normal for the month, temperatures above.

On Work Songs: Singing It In

Max Godfrey

Before you can use a work song in a work situation, you have to get to know it for yourself. You have to "sing it in," as traditional singer Heather Wood would say, before you can teach it to others. Your brain may have memorized the words and the melody, but the brain is nowhere for a work song to live. When you've been harvesting all morning and find yourself out in the field at 3 pm, dog-tired and starving for lunch but still trying to transplant a round of seedlings before the afternoon rainstorm blows in, your brain isn't much good for anything. Since the moments when work songs are the most important are the moments when the your head and your muscles have started to shut down, a work song must reside in some place deeper within yourself. You have to make a home for the work song in your gut.

In Pine Mountain, Georgia I would often walk out into a neighboring cornfield and sing before work began. And most nights I made a point of singing a couple songs outside my trailer to put myself to sleep. I would sing very slowly during these moments, sometimes holding out particular notes for a long time, allowing those notes to resonate throughout my body, especially in my belly and my jawbone. I've found that even highly rhythmic songs, such as Berta Berta, take on a rich, and even captivatingly lonesome character when they are sung with a slower, more spontaneous rhythm. I leave enough time between lines to hear my voice echoing off the forest at the edge of the field. By slowing the songs down, I've found myself spending ten to twenty minutes running through a work song, settling into a meditative state as I go along. By singing in solitude, free to make mistakes and improvise, we can take traditional songs we've learned from recordings and make them our own.

Most of the songs I sing I have learned while I've been sitting indoors, with my ears pressed up against a computer speaker. The same, I believe, is true for many of today's traditional singers. Most of us haven't grown up with these songs. We have no faces, no scenes, no smells or memories to connect them to. But it is the singer's own experiences with a particular song, rather than the melody and lyrics alone, which really bring a song to life and make others want to join in. For this reason it is all the more important that we spend time with our songs: take walks and measure our steps with them, stop and sing them at sunrise, or while you're coming back home from the field, or while you're doing the dishes, or staring into the wood stove in the evening. Get to where they mean something to you. Share them with one or two people at a time and see how they respond. Watch how the rhythm wants to change as they join in.

Eventually a song will start letting you know that it wants to be sung. Listen to that impulse and follow it, if it means getting out of bed in the middle of the night and stepping out into the cold to sing, or even if it means bursting into song on a sidewalk as strangers walk by. If you obey these impulses, you'll find that you are changing yourself to fit a song more than you are changing a song to fit yourself. A song is actually pushing you outside of your comfort zone. But eventually you'll start to feel at home with it—grounded, focused. And when the week has been too long, the days too wet, too cold, when you're feeling lonely or discouraged, a song will come welling up from your gut

and bring peace to your mind. After singing it through you might even be able to laugh at your troubles.

The sense of peace you feel with a song will come through in your singing, and your fellow workers will draw from it as they join in. Just remember that before you can use a song as a tool for creating balance within the work crew, you must be singing it from a place of balance. And more importantly, for a song to bring joy to your fellow workers, you must be singing it from a place

of joy. So find a song that demands something from you. One that tells you to stop right where you're at and to give some of your self to it, to make some time for it in your day and to make some room for it in your gut.

And then offer it more. Keep singing and see what it gives back to you.[1]

---

[1] See the songbook at the end of this volume for a collection of work-songs.

THRESHING SONG

Flatbread Society
www.flatbreadsociety.net

This music is open source.
Share freely.

MMXIV

Written & composed by:
Andy Cabic & Otto Hauser

Debut:
Eli & Edythe Broad Art Museum
Michigan State University

The Land Grant: Art,
Agriculture & Sustainability

Fig. 133.—MORRISON WAR-TIME BODY WITH NO PANELS

Fig. 134.—MORRISON WAR-TIME BODY WITH CENTRE OPENING AND SHUTTERS AT REAR END

Fig. 34.—AN 8/10-CWT, MURPHY ELECTRIC
With body designed for the delivery of dairy produce.

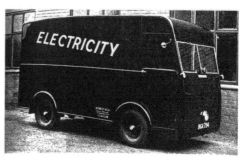

Fig. 56.—A VICTOR ELECTRIC 2-TON DELIVERY VEHICLE

Fig. 33.—Partridge Wilson electric.
With tipping body for delivery of coal.

Fig. 57.—A streamlined "Metrovick," electric for light showroom deliveries.

Fig. 42.—A G.V. electric
With dustproof tipping body for refuse collection.

Fig. 41.—50-cwt. Morrison-Electricar van

EDITORS NOTE Before the technology tree grew in the direction of gasoline, it was neck and neck for a moment with electric and steam cars. There's a whole hidden history of this part of the relic trajectory, at the Prelinger Library in San Francisco. Open on Wednesday afternoons. www.prelingerlibrary.org

Ag's White Truck

Heidi Hermann

The presence of a white truck in the agricultural landscape is virtually a given. Across the nation this phenomenon exists. I have worked in various echelons of agriculture (ag), from research to governmental to management to field hand. At all these stages there was the obligatory white truck on the scene. I've seen them creep into the field from afar, I've driven them, and I've gotten rides in them. They seem to equate the driver with some sort of leadership position. They are definitely noticeable from a distance; the white metal is a stark contrast to the browns and greens of the landscape. It invokes the feeling: "Oh, here he comes" in the minds of the field hands.

Not only is the white truck ubiquitous with a sense of power on the agrarian landscape, but so is the notion that the driver will be a man. To his industry peers he is recognized as a man with a purpose and responsibility on the job; he's been trusted with the company vehicle and he's keeping it clean. For example, I took conducted an informal

DRIVING FOR PLEASURE

study in Salinas, CA on who drives these things. A non-biased discovery was that white, young (25-40), clean-cut guys were the predominant drivers. What gives? White guy = white truck? There was that 3% though, of Latinos or women who drove them. Perhaps these are folks who had worked their way up the ladder to crew manager and were given 'loftier' responsibilities. Also

worth noting as a young woman driving a white truck in Salinas at the time, I got many surprised (or were they flirtatious?) looks from men. "There is somebody different from us on the scene?" "What's she doing?" "Who's she working for?" I could imagine onlookers asking themselves.

Why Trucks?

Trucks are the old standard for farmers. Are the people that drive these things truly farmers though? Generally, they are at least in the industry if they roam into the fields. But rarely do the drivers plant the crop or do the actual hands-on farming. These white trucks belong to the management or related echelon, i.e. PCA, researcher/extension agent, fertilizer salesman, etc. Field and crew managers generally get issued one of these icons if working for a larger establishment, or purchase them on their own accord. Team unity through identical company vehicles is good in theory—it creates uniformity, homogeny and professionalism. These are the guys that drive from site to site hurriedly, hopping out now and again to chat with one of the crew members, relay messages, or unload/pick-up supplies. Drivers are often labeled with the term "tan-left-arm managers." You've seen these guys—their arm resting out the driver window, busily puttering about, multitasking, eating, conducting business from their mobile office on the front seat all while perusing muddy and uneven ranch roads. They have work clothes on, but are unsoiled. Trucks are designed to carry a cargo in the bed, but given today's common farm-sized acreage, this little 5'x6' just won't suffice, so it usually goes empty and its utility is diminished. The truck is not used for its initial design purpose, but rather a vestige of its former function.

Why white?

It seems rather impractical that farm trucks are commonly white since they're driven on dusty or muddy fields. However, you rarely see these vehicles splattered in brown schmeck. They are generally kept quite white. But why and how? White represents clean, and with a crisp, newly-glistening, washed truck the driver appears new, fresh, clean and honest. For this to invoke trustworthiness from such an untainted, virgin surface may be a reach, but it happens all the same.

It is inevitable that mud gets splattered on these unmarred surfaces at some point in the day. But, somehow by tomorrow it shows up clean again. Who cleans these things? My wager is some other 'lower echelon' staff person does the cleaning-up at the end of the day. Or as my employer advised, "Just bring it down to any of the car washers in town." They were always well attended. A sign that repeatedly humored me as I waited in line behind other soiled white trucks read: "$5 extra for dangling mud." This evidently muddied-up their scene, too.

Up—up—up has gone the cost of living for the horse. And down—down—down has come the price of the easily-operated and economical Ford, to a point where no farmer can afford to keep a horse for road travel only.

"In society today, technology companies can do what they damn well please, but this is the Grange. This is a fraternity. Attorneys have a different sense of things, that's their teaching and training. This type of organization needs to work for the good of all. Be mindful of the younger people coming in, they are comfortable with technology. I was in technology before I got into poultry, and I like turkeys better than programmers."

—From meeting of
California State Grange

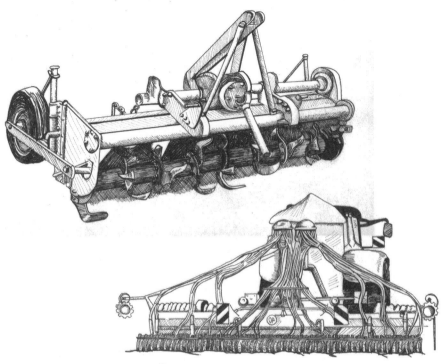

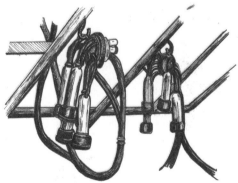

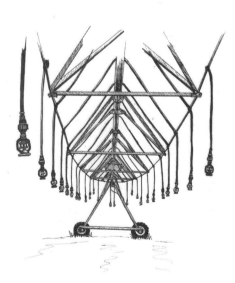

Comparison of home garden tools to their agricultural equivalent. Our tools may differ but we all work the ground!

Illustrations by Erin Smith
Concept by Jeff Hake

Douglas Rainsford Tompkins

Douglas Rainsford Tompkins

EDITOR'S NOTE From the air the interactions between farmers and the contours of the land become more apparent, and add further depth to our understanding of these relationships. So if you have access to a plane, use it!

Alex MacLean

Alex MacLean

Interview with Eric Herm
Conducted 12.6.12 in Brooklyn, NY

Helen Zuman

After planting Monsanto's Roundup Ready cotton seed on part of his family's land—and not getting the higher yields he'd expected—Eric Herm swore off GMOs, weaned the entire operation off chemical pesticides and fertilizers, and began to experiment with growing cotton organically. He now advocates for farming practices that treat nature as a partner, not an adversary.

HELEN ZUMAN You're a fourth-generation farmer. Your family has been farming in Texas for almost a century. When and why did your family start using chemicals for fertility and pest control? Did your father or grandfather have doubts about adopting and using the new methods? Have they ever expressed regret over their switch to industrial agriculture?

ERIC HERM No, they've never said they regretted it, I think because it was so gradual. When they first started having problems with weeds, I don't think a lot of farmers understood why weeds grow, why insects were destroying their plants. They were just looking for quick fixes. I think education and lack of awareness is our main problem in agriculture. A lot of young men who started farming in the early 1900s didn't really know much. They were just looking for a good, simple life. They didn't get a whole lot of schooling in farming. You learned what your dad knew, what those around you knew. Everybody knew the same things 'cause they

were all in it together. Nobody was fulfilling that role of teacher. In stepped the corporations, in the early 1900s. These companies—the Monsantos and the DuPonts—were war companies. They were making bombs and bullets and chemicals for war. After World War I, they had all this stuff left over and they sold it to farmers! And the farmers were like yeah, okay, we'll use it, it makes the insects go away. This is great, this is magic! They refused to ask—can this stuff hurt us? Is it bad for me? Is it bad for the land?

ZUMAN What exactly do GM crops do to the soil, the ecosystem, the bodies of humans and animals who live near them and ingest them?

HERM They degrade the soil. We see superweeds developing in our neighbors' fields. They can't kill them with Roundup anymore because they've become resistant. Looking at fields that have been planted in Roundup Ready crops, year after year, is like looking at a sick old person who's hooked up to an oxygen tank and still smoking cigarettes. No one seems to make the connection. That's what we're dealing with now. And now they're coming out with new poisons and new seeds. Seed that's resistant to 2, 4-D is the next strain they're gonna push on the market in a couple years. That's gonna be even more destructive, 'cause it's such an aggressive herbicide. It kills all broadleaf plants. We've already experienced problems with Roundup drift from our neighbors. It made our plants sick.

They wouldn't grow for three weeks. Drift from 2, 4-D would completely kill our plants. It's that much more aggressive.

ZUMAN: You no longer use genetically modified seed of any kind. What convinced you to stop? Describe the moment you realized your experiment with genetically modified cotton had failed.

HERM Well, I don't know that it failed. It was just a conscious choice. We'd only done it two years, and we didn't see that it was making higher yields for us. We looked at it from a business point of view. It wasn't increasing our yields. But it was a consciousness that caused us to stop. I look at that decision now—it's been six years since we made that decision—I look at our soil, and even though we've been in a two-year major drought, our fields are in better shape than ever. We have less weeds than ever. We have less insect issues than ever. It just verifies that decision that we made, how much healthier our soil is and how much healthier our crops will be, in the future.

ZUMAN You recently had a problem with drift from Monsanto cotton.

HERM Yes. We had probably three hundred acres on four different fields that suffered from drift from Roundup from our neighbors. Like I said it doesn't kill the plant—it just won't grow for three or four weeks. It's basically trying to process all the poison out of its system, before it can kick in and start being a plant again. I filed reports with the Texas Department of Agriculture. When I did that I became very unpopular. About fifteen different neighbors had to submit their spray reports to the Texas Department of Agriculture to show when they sprayed, what day, what time, the conditions and all that. I made them mad! But they weren't near as mad as I was. They were killing my plants, or weakening them. My organic field is two hundred and fifty acres and probably fifty to sixty acres of it was hit. It killed some of those plants

'cause I'd just planted them. They'd been above the soil for less than forty-eight hours, they were just babies, just seeing sunshine for the first time. My logic was that I've gotta reach these guys somehow, because they're hurting us, they're preventing us from farming in a healthy way. What they do impacts us. I think I got my point across. People talk–they don't tell me directly, but you hear things. At least they know now that I'm willing to fight back and not just say, hey, please watch out. I'm willing to take the next step. The sad part is, you're looking at neighbors who are gonna have to get into court cases with each other, because some guys don't want to 'fess up that they're the ones who sprayed. They're gonna blame the guy next to them. That's why I get the TDA involved. They can police it, they can find out who did what on what day.

ZUMAN I'm wondering about what makes you different. You traveled down the path of using chemicals and GMOs, and then you came back, because of a change of mind, a change of heart. Is that change replicable? What would it take for others—farmers who have taken that road, who are doing industrial agriculture—to do the same? To come back, like you did? Maybe a lawsuit is part of it, maybe some regulatory authority coming down on them is part of it—but what would it take for them to feel the need to do something different?

HERM For a lot of guys it's going to have to be economical. Economics is unfortunately the major motivator for a lot of people. The sad part is, most farmers want to be good stewards of the land, they want to be. But we've been brainstained to do things the easy way. And a lot of farmers are resentful because they were ignored for so long. Nobody cared. Farmers were suffering a lot in the 80s and 90s. You look at the American Agriculture Movement, which happened in the late 70s—nothing really happened as a result of that. A lot of these guys fought, they sacrificed a lot, and they didn't see any benefit from it.

ZUMAN Some say we need chemicals and genetically modified seed to "feed the world." That we're being elitist and impractical if we don't back these technologies. What do you think of this argument? How successful has chemical farming been, to date, at feeding the world, especially the poor?

HERM Do you want to feed the world, or do you want to nourish the world? You look at everything from livestock to ourselves, and yeah, we got plenty of food out there, but the quality is bad, because we've got sick soil. Our soil's been bombarded with chemicals and synthetic fertilizer. We're not growing healthy crops, for the most part, so we're not gonna be healthy people. That's just one of their gimmicks, it's one of their sales pitches that they use to reach most people. Somebody can pick up a magazine and see that and be like, "Yeah! Feed the world! That's what we gotta do!" We don't take the time to ask, how are we going about that?

ZUMAN What would it take to nourish the world?

HERM The desire. We need to start looking at crops as seed, as soil, as an extension of ourselves, as life, as energy, instead of a dollar sign or a commodity. We need to start making that connection back to nature, instead of seeing eating as something we do in between stop lights.

ZUMAN California's Right to Know proposition, which would have required food companies to list genetically modified ingredients on product labels, was recently defeated despite initial polls showing overwhelming support. One argument Monsanto and company used to defeat the measure was that it would raise the average family's grocery bill by $400 per year. Whether or not this argument is true, it seems to have resonated with the public. Why do you think Americans are so obsessed with paying as little as possible for food?

HERM I have no idea. It makes no sense. People want to spend as little as possible on their food, but we'll go out and buy a two-hundred-dollar iPhone, or the latest, greatest contrivance or device. I don't think people make the connection between how much more we're spending on health care, or doctors and pills, and the weaknesses and illnesses we get from inferior food. Again, it's that quantity-over-quality mentality. We don't seem to make the connection: Food becomes us. What you eat can enhance your energy or it can weaken you. We seem to be disturbingly content with the latter, because it costs less. We don't seem to make the long-term connection that it's gonna really bite us in the booty later.

Make a Portable Planetarium
By Fritz Horstman

Find an empty container,
such as a salt cylinder or tissue box.

With a thumbtack or small nail,
poke holes in the container. Remember
that the pattern you make will be the
opposite of what is seen from the inside.
You can make recognizable constellations,
or invent your own!

Hold the opening of the container to your eyes.
Do you see the pattern of holes that you made?
Sometimes gluing some black felt around the
eyehole helps to block unwanted light.

Pat Perry *Great Lakes Tar Sands Resistance*

CUBE COLA

STEP 1: THE EMULSION FORMULA

Use food-grade essential oils only.

7.50 ml orange oil
7.00 ml lime oil
2.00 ml lemon oil
0.75 ml cassia oil
1.50 ml nutmeg oil
0.50 ml coriander oil (12 drops)
0.50 ml lavender oil (12 drops)

Using a measuring syringe, measure out the oils into a glass or ceramic container. Keep the oils covered to avoid volatile fumes escaping.

20 g food grade freeze dried gum arabic (equivalent to 44ml)
40 ml water (low calcium / low magnesium)

Use a pestle and mortar to dissolve the gum arabic into the water (with optional 1 drop vodka which can aid hydrofelicity, the total quantity of vodka will be 0.0003ml per litre of cube-cola).

Place the gum/water mix in a high-sided beaker, Pyrex glass or stainless steel are best. Use a high-power drill (the greater the RPM the better) with a hand kitchen whisk attached; you may need to modify the handle of the whisk so it fits. Whisk the gum mixture at high speed while your colleague droppers the oils in steadily with the measuring syringe. Continue to whisk at high speed for several minutes, or until you can see the oils and water emulsify.

The resulting mixture will be cloudy. Test for emulsification by adding a few drops of the mixture to a glass of water. No oils should be visible on the surface. You now have a successful flavour emulsion, which should hold for a minimum of 12 months.

TOTAL FLAVOUR EMULSION MEASUREMENT: 65ML

If emulsification is successful, continue.

STEP 2: THE CONCENTRATE

195 ml double strength caramel colouring
(DD Williamson Caramel #050)

Add the caramel colouring to your 65ml emulsion formula.

Then make a caffeine/citric solution:

65 ml citric acid (use a metric measuring spoon to measure)
100ml water
18ml caffeine (powder)

In a mortar and pestle, mix the citric acid into the water. This should dissolve easily. When it's clear, use a sieve to add the caffeine. Mix thoroughly, the caffeine will take 5-10 minutes to dissolve. The mixture may behave erratically, turning either white or clear for no apparent reason.

Add the citric/caffeine solution to the emulsion formula, passing it through muslin or jelly bag to remove any anomalies. This is your cola concentrate (total measurement 365ml / equivalent to 90L eventual cola).

To subdivide the concentrate into manageable quantities, please consult the following table. We do so for easy mailout and long-distance production into cola syrup with the local addition of sugar and water.

TABLE OF CUBE-COLA VALUES

Large _ 56ml makes 14ltr Cola
Small _ 28ml makes 7ltr Cola

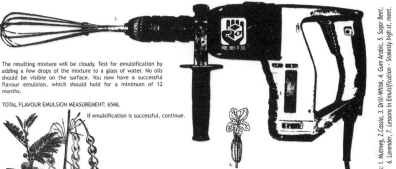

Images: 1. Nutmeg, 2. Cassia, 3. Drill-Whisk, 4. Gum Arabic, 5. Sugar Beet, 6. Lavender, 7. Lessons in Emulsification - Stokesly high st, meet.

STEP 3: THE COLA SYRUP

750ml water
(we tend to use filtered tap water)
1.5KG granulated sugar
(beet sugar if available)
56ml Cube-Cola concentrate

Make a sugar syrup (mix in a cooking pot on low heat to speed dissolve) with 700ml of the water and all of the sugar.

When the sugar syrup has cooled, mix in the cola concentrate, using the remaining 50ml water to rinse out the concentrate jar to ensure full inclusion. You now have approximately 1.75L Cube-Cola syrup or an eventual 14L Cube-Cola.

STEP 4: THE COLA

As required, make up your cola as a 7:1 mix, 7 parts carbonated water to 1 part cola syrup. We currently use 250 ml syrup in a 2L bottle of sparkly water, the cola can alternately be composed direct in the drinker's glass.

NOTES:

1 batch of this recipe as written here will produce approximately 90 L cola. You may want to divide all quantities by 1/2 if you do not want to become a small cola producing factory; smaller divisions of the recipe are difficult to handle at the emulsion stage, as the measures of oils and water are already very slight.

Cube-Cola est.2003 manufactured inhouse at the
CUBE MICROPLEX, BRISTOL
4 Princess Row, Bristol BS2 8NQ, UK
By Kate Rich and Kayle Brandon
cube-cola.org
cola@scubecinema.com

STANDING ON THE HANDS OF GIANTS

Kate Rich & Kayle Brandon

September

WHEN LIFE GIVES YOU APPLES MAKE CIDER

Democracy

Chapter 9

Wheat Mall:
Agriculture Public Art

William Giordano

Wheat Mall was a public artwork that took place in Orono, ME during the 2012 growing season. The goal of the project was to promote food production in the commons. The project was completed in three steps:

1. Design and planning

2. A campaign aimed at winning over those who could prevent the work

3. Crop production, celebratory work events and an installation

The work exalted the legacies of agriculture and landscape architecture in the State of Maine dating back to the early to mid-1800's. The two historic figureheads whose work made *Wheat Mall* possible were Ezekiel Holmes and Frederick Law Olmsted.

The project team attained necessary permissions and plowed the Nationally Recognized Historic Lawn at the University's Orono campus to grow "Red fife", a heritage wheat. The crop was grown on four plots of reclaimed lawn in high visibility areas of the University of Maine's campus totalling approximately 8,000 square feet. The land was tilled and the crop planted mechanically, and all subsequent cultivation, harvesting and processing were completed with simple hand tools. The project team worked closely with the crew at the University's Grounds Shop, research farm, and the general public. The crop was harvested with hand scythes and processed

William Giordano *Wheat Mall gallery installation*

into flour by the community. Processing took place during celebratory threshing and winnowing events and a month long gallery installation that invited viewers to ride a bike-powered grist mill. The flour was donated to the distinguished regional old-world bakery Tinder Hearth, in Brooksville, ME.

Wheat Mall set out to produce food in common and historically-valued space. Equally, however, it was an experimental campaign targeting land use planning committees and University administrators. In fact one large plot of wheat was cultivated in back yard of the President's on-campus home. The campaign successfully won over the audience, who then granted permission and support for the work. The primary tactic of the campaign was accessing a positive pastoral cultural

William Giordano *Wheat Mall plowing and planting*

memory that is connected to relevant agricultural and food system challenges of today.

Statement of Purpose

The way we engage with land and landscape is an expression of our cultural ecology and our economic values. The ecologically overburdened landscape, symptomatic of an extractive and exploitative economy, is troublesome. Yet it is an often-ignored reality in American cities, towns and villages. It is critical to create broad and diverse positive responses to the economic and ecological challenges of our time. This project employed agriculture and public art to deliver that message. Over the past century we have seen a significant shift in public landscapes away from agriculture, towards recreation and private development.

Today, Maine's food system is faced with some of the most significant economic and ecological challenges we have seen. Publicly owned lands are a ripe canvas for these positive responses to take place.

Wheat Mall combined the disciplines of agriculture, history and public art to create a meaningful project, which encouraged citizens to consider their relationship to land, food and landscape. Wheat Mall created a functional multi-use landscape that allowed the community to engage the prominence of agriculture upon it. Furthermore, the project invited a large audience with diverse food buying habits to inquire deeper regarding into our relationship to food and local land.

William Giordano *Wheat Mall plowing and planting*

William Giordano *Wheat Mall harvest*

Historic Research

Wheat Mall was historically shaped by the legacy of two vital figures in Maine History, Frederick Law Olmsted, Sr. and Ezekiel Holmes. Their contributions to land use and Maine's agriculture respectively, are on display at the University of Maine. The influences of Olmsted's 1863 design for the campus and Holmes' pioneering agricultural leadership provide a historical foundation for *Wheat Mall*. Furthermore, the project connected its historic relevance with contemporary agronomy via Northeast Local Bread Wheat Research Project.[1]

The *Wheat Mall* project took place within the National Register Historic Area of the University's Orono Campus. Olmsted's design is part of the reason for the National Historic classification. His 1863 campus plan called for experimental crops to be grown by students in the areas that *Wheat Mall* took place.[2] Ezekiel Holmes was a Land Grant pioneer whose work led to the founding of Maine's State Agricultural College (now the University of Maine). He farmed in Winthrop and travelled extensively to learn about and return home with seed and livestock

he thought would become assets to Maine's people. He published the original monthly newsletter, "The Maine Farmer" from 1833 until 1865, and called for the ongoing experimentation and development of crops and livestock on small and large scales across Maine.[3] Holmes believed that the ongoing development of a yearly wheat crop, even amongst the drastic changes in agriculture and economy during the second half the 19th century, would be essential to maintain a healthy State.[4]

1   Amaral, Jim; Bramble, Tod; Mallory, Ellen; Williams, Matt. *Understanding Wheat Quality: What Bakers and Millers Need and What Farmers Can Do.* University of Maine Cooperative Extension. Bulletin #1019. 2012: Orono, ME.
2   Olmsted, Frederick L. *Architect's Report to the Board of Trustees of the Maine State College of Agriculture.* Annual Report, State of Maine, 46th Legislature. House, no 87. p. 31

3   Day, Clarence A. *Ezekiel Holmes; Father of Maine Agriculture.* University of Maine Press. 1968: Orono, ME.
4   Holmes, Ezekiel (ed). *Maine Farmer & Journal of the Useful Arts.* Vol. 1, No. 14 & Vol. 2, No. 11 . William Noyes and Co. 1833–1834: Winthrop, ME.

Coöperative Finance

An American Method *for the* American People

TO encourage Business, Farming, Home-owning, Individual and Corporate Success, Social Justice and National Prosperity.

How to re-form banking, improve the mechanism of exchange, promote co-operation, and insure against the possibility of a money trust, through combining American experience with scientific principles, by means readily adapted to customs, conditions and institutions throughout the United States.

By HERBERT MYRICK

Profusely Illustrated by Charts, Designs and Good-Humored Cartoons

An American Method of Co-operative Finance *to*

1 — **Wisely Mobilize the Bases of Wealth** — gold and effort, land and industries, labor and character, health and co-operation — as a means for credits and exchanges through a sound banking system

2 — **Safely Furnish Suitable Banking Facilities** — to all the people all the time, and ever be adequate to the country's needs for expansion or contraction

3 — **Supply Rediscounts** for commercial paper and farmers' notes, "acceptances," bills of exchange, "clearings," collections, national reserves, credit currency in addition to government money, etc., thus remedying all the defects in our present banking system while making the most of its advantages

4 — **Insure That All Deposits Shall Be Available** when wanted, furnish giltedge investments, "accommodate" farmers, other workers and small businesses, as well as rich individuals and larger corporations

5 — **Provide Land Banks to Issue Non-Taxable Bonds** — secured by mortgages upon real estate — peferably owned farms and homes, thus encouraging agriculture and home-owning, while enabling other banks to realize on such mortgages

6 — **One Comprehensive Act of Congress** to establish FORTHWITH this American monetary method, to be supplemented in due course by appropriate legislation in the respective states, in the interest of productive industry rather than of stock exchange gambling

7 — **This Method Will Insure Stability in Finance**, facilitate exchange, reduce expense, risk, and loss, be reasonably profitable, increasingly efficient, beneficial to all, harmful to none, promoting domestic industry and foreign trade — the foundation of an economic prosperity that shall foster social justice, and make the United States more than ever the hope of the world

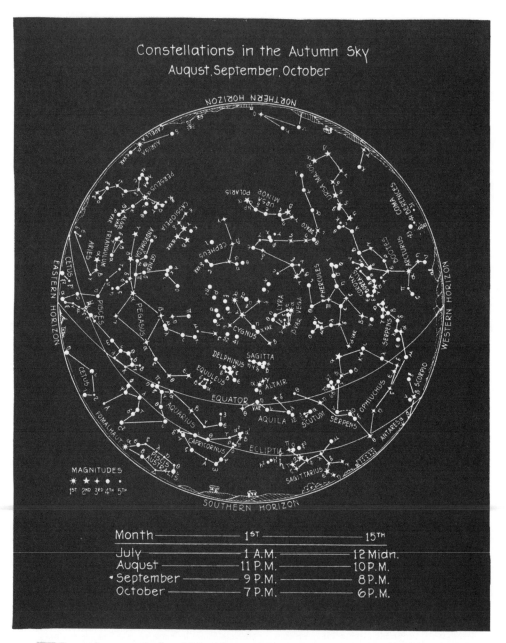

Constellations in the Autumn Sky
August, September, October

Month	1ST	15TH
July	1 A.M.	12 Midn.
August	11 P.M.	10 P.M.
September	9 P.M.	8 P.M.
October	7 P.M.	6 P.M.

TO understand the "Northern Horizon" on the map, the observer should face north and hold the chart so that the words "Northern Horizon" are directly below the center of the map. Do likewise for stars in the south, or west. One may well learn the most striking constellations first.

CORRECTIONS TO THE TIMES OF SUNRISE AND SUNSET

On account of the use of standard time it is not possible to give a table of sunrise and sunset applicable to every city without error, unless a special table is provided for that place. The error may amount to three-quarters of an hour, fast or slow, and for those wishing more accurate times of sunrise and sunset the following table of corrections is provided. Add or subtract the number of minutes indicated in the table.

Akron, O.	add	26m		Los Angeles, Calif.	subtract	07m
Atlanta, Ga. E.T.	add	38m		Louisville, Ky.	subtract	17m
Atlanta, Ga. C.T.	subtract	22m		Memphis, Tenn.		00m
Albuquerque, N. M.	add	06m		Milwaukee, Wis.	subtract	08m
Baltimore, Md.	add	06m		Minneapolis, Minn.	add	13m
Bay City, Mich.	subtract	24m		New Orleans, La.		00m
Bismark, N. D. C.T.	add	43m		New York City	subtract	04m
Bismark, N. D. M.T.	subtract	17m		Oklahoma City, Okla.	add	30m
Boise, Idaho	add	15m		Omaha, Nebr.	add	24m
Boston, Mass.	subtract	16m		Peoria, Ill.	subtract	01m
Buffalo, N. Y.	add	15m		Philadelphia, Pa.	add	01m
Canton, O.	add	26m		Phoenix, Ariz.	add	28m
Charleston, S. C.	add	19m		Pierre, S. D. C.T.	add	41m
Cheyenne, Wyo.	subtract	01m		Pierre, S. D. M.T.	subtract	19m
Cincinnati, O. E.T.	add	38m		Pittsburgh, Pa.	add	01m
Cleveland, O.	add	27m		Portland, Ore.	add	11m
Dallas, Tex.	add	27m		Providence, R. I.	subtract	14m
Dayton, O.	subtract	23m		Richmond, Va.	add	10m
Denver, Colo.		00m		Saginaw, Mich.	subtract	24m
Des Moines, Ia.	add	14m		Salt Lake City, M.T.	subtract	28m
Detroit, Mich.	subtract	28m		Salt Lake City, P.T.	subtract	32m
Duluth, Minn.	add	08m		San Antonio, Tex.	add	34m
Flint, Mich.	subtract	25m		San Francisco, Calif.	add	10m
Galveston, Tex.	add	19m		Spokane, Wash.	subtract	10m
Helena, Mont. M.T.	add	28m		St. Louis, Mo.	add	01m
Helena, Mont. P.T.	subtract	32m		St. Paul, Minn.	add	12m
Houston, Tex.	add	21m		Springfield, Ill.	subtract	02m
Indianapolis, Ind.	subtract	15m		Tampa, Fla.	add	30m
Kansas City, Mo.	add	18m		Topeka, Kan.	add	23m
Lansing, Mich.	subtract	22m		Utica, N. Y.	add	01m
Little Rock, Ark.	add	09m		Washington, D. C.	add	08m
				Youngstown, O.	add	23m

Put Second Bill of Rights in Constitution

In our day these economic truths have become accepted as self-evident. We have accepted, so to speak, a second Bill of Rights under which a new basis of security and prosperity can be established for all—regardless of station, race, or creed.

Among these are:

The right to a useful and remunerative job in the industries, or shops or farms or mines of the Nation;

The right to earn enough to provide adequate food and clothing and recreation;

The right of every farmer to raise and sell his products at a return which will give him and his family a decent living;

The right of every businessman, large and small, to trade in an atmosphere of freedom from unfair competition and domination by monopolies at home or abroad;

The right of every family to a decent home;

The right to adequate medical care and the opportunity to achieve and enjoy good health;

The right to adequate protection from the economic fears of old age, sickness, accident, and unemployment;

The right to a good education.

All of these rights spell security. And after this war is won, we must be prepared to move forward, in the implementation of these rights, to new goals of human happiness and well-being.

America's own rightful place in the world depends in large part upon how fully these and similar rights have been carried into practice for our citizens. For unless there is security here at home there cannot be lasting peace in the world.

SUBJECT: ALMOST ANY POLITICIAN; SCENE: ALMOST ANY COUNTY SEAT TOWN

John Jimson was a lawyer in a little country town; he was good at pettifoging with a nose just like a hound on the scent of petty quarrels which would bring him juicy fees; nothing was too low or sordid when his avarice would appease. It was fun to see him swagger; how this shyster would cavort; he would have a dozen cases in each session of the court; he would bullyrag a witness who he knew could not resist; he would thunder at the jury 'till their thoughts were all atwist. He was crooked in his dealing; he would steal his clients blind; and a weeping trusting widow, was to him, a lucky find; if her husband left insurance, it would always find its way to this swindler's greedy pockets; and in spite of tears, would stay. Now this fellow got a notion that in congress he would shine; though God knows, his brains were minus and his head was knotty pine; so he strutted round the country kissing voters' dirty kids, and for honest farmers' ballots, he was lavish in his bids. Though this guy was very crooked and all knew him for a skate, when he did his yearly voting then his politics were straight; so the party got behind him and to congress he was sent; for the farmers worshiped party and their votes could not be bent. Now he takes his orders meekly from the bosses of the land; he was just the tool they needed to support their robber band; and though farmers grunt and grumble, they are just a bunch of goats, for this fellow was elected by the darn fool farmers' votes.

A. M. Kinney.

"Certain persons have driven a herd of cows, on whose milk they live, into an enclosure. The cows have eaten and trampled the forage, they have chewed each others' tails, and they low and moan, seeking to get out. But the very men who live on the milk of these cows have set around the enclosure plantations of mint, they have cultivated flowers, laid out a race-course, a park, and a lawn-tennis ground, and they do not let out the cows lest they should spoil these arrangements. ...The cows get thin. Then the men think that the cows may cease to yield milk, and they invent various means for improving the condition of the cows. They build sheds over them, they gild their horns, they alter the hour of milking, they concern themselves with the treatment of old and invalid cows ... but they will not do the one thing needful, is to remove the barrier and let the cows have access to the pasture."

—Leo Tolstoy, *A Great Iniquity*

NEW ENGLAND WANTS SPEED
NORTH EAST WEST SOUTH

DO YOU KNOW

THAT ALTHOUGH BOSTON has a more accessible and commodious harbor, less fog, and is a day's sail nearer Europe than New York, even New England exporters use the port of New York because the delays and expense of the competitive railroad system at Boston consume the entire value of the natural advantages?

THAT SHIPS BRINGING cargoes for, or receiving cargoes from any railroad in Greater Boston should use the docks where that railroad has its terminals, but that no dock is directly accessible to more than one of the three railroads which serve Greater Boston?

THAT SHIPPERS have to pay a heavy switching charge each time goods in less than carload lots or from points East of the Hudson River are transferred from one railroad to another in Greater Boston?

THAT FREIGHT AND PASSENGER rates for equal distances vary according to which railroad is used and what point in Greater Boston the freight or passenger is destined for or from?

THAT FREIGHT DESTINED for most parts of the Greater Boston area is shipped downtown for distribution by truck from the vicinity of the North or South Terminals?

THAT MUCH OF THE TRUCKING in downtown Boston is caused by the hauling of freight between the North and South Terminals, between railroads and docks, between docks and docks, between the North Terminals and factories and warehouses in the southern part of Greater Boston and between the South Terminals and factories and warehouses in the northern part of Greater Boston?

THAT IN MAKING a shipment from Lynn to Hyde Park, a distance of about 25 miles by existing lines, the shipment travels from Lynn to Salem, Salem to Lawrence, Lawrence to Lowell, Lowell to Mansfield, Mansfield to Readville, Readville to Dedham or Hyde Park, a distance of over 100 miles and requires 6 days in transit?

THAT WHILE RAILROADS are losing local passenger patronage, laying off trains and boarding up stations, the Legislature is providing for the building of new rapid transit extensions of the Boston Elevated system, duplicating trackage already built and owned by the steam roads, at a fabulous cost, to be borne in large measure by the taxpayers?

THAT GREATER BOSTON is almost the only remaining great metropolitan area in the country where no attempt has been made or is likely to be made by the standard gauge railroads to electrify their lines?

THAT SEVEN DIFFERENT railroad and street railway companies operate in Greater Boston, all with varying rates of transportation per mile, and all requiring the payment of additional fares when it becomes necessary to transfer from one to the other?

THAT ALL THIS is costing passengers, shippers, manufacturers, consumers and the general public of New England many millions of unnecessary dollars each month?

At the top we look into a refrigerator car. One man is rolling a sheep and the other a quarter of beef into the cold-storage house. Below are firkins of butter going into the cold-storage house.

The Panhandle of Idaho

Has Greatly Diversified Land, climate, altitude and products. It is a country to suit all comers. On one farm of 160 acres was grown a total of 112 different varieties of products!

IDAHO and the NORTHERN PACIFIC

NORTHERN PACIFIC
YELLOWSTONE PARK LINE

"Together [the] railroads formed a lever that in less than a generation turned western North America on its axis so that what had largely moved north-south now moved east-west. Railroads poured non-indigenous settlers into a vast region that nation-states had earlier merely claimed. They did not do this in response to a popular demand for development of these lands; instead, they created the demand through vast promotions unlike anything seen until that time. Having promoted new settlement, they helped integrate these settlers into an expanding world economy so that wheat, silver, gold, timber, coal, corn, and livestock poured out of it."
—Richard White, from *Railroaded*

DOWN TO GRASS ROOTS

Minnesota

has millions of acres still awaiting the right kind of men to develop them and make them productive.

MINNESOTA and the NORTHERN PACIFIC

NORTHERN PACIFIC
YELLOWSTONE PARK LINE

Northern Pacific Railway

SEASONABLE SUBJECTS

WASHINGTON and the NORTHERN PACIFIC:

Washington

Land of snow-capped peaks, noble forests, tumbling rivers, fertile valleys, wave-lapped seashore. Incomparable climate. Immense water power; vast areas of valuable timber; a fishing industry that rivals the world. Intensive irrigated farming and fruit-growing in a high degree of development. Dairying and stock-raising on extensive scales. Every thing and every man in Washington is making money!

Northern Pacific Railway

Revelation

Savitri D

Were you watching television in a motel in Bakersfield? Staring at the reflection of police sirens in the wet streets of Providence? Listening to a casino boat rock against the banks of the Mississippi? Were you under your grandmother's table? Lacing up your skates at the community center? Were you watching the curtain sway before the lights went down?

Ask a radical how they were radicalized. Was it a gradual undoing? A trauma witnessed? A story? An existential crisis? An association? An education?

We come to our radical lives from many directions, but eventually the radical life IS the direction. And we grope and we fail. Rewards are short-lived—the condition of radicalism precarious. We bolster ourselves with the hope of history but that is untrustworthy. We gather in clusters around identity, issues, strategies.

We are corralled into careers, forced to limit our tactics to satisfy the requirements of funders. We are still groping and still failing. Bored by the platforms of the modern moment, we are cornered by security problems, continual surveillance, hacked email, the inconvenience of legal battles. We are lucky to have food, shelter, books, lawyers and most of all we are

Dorothy Day

lucky to be artists. That too is a precarious life. Like the great Japanese Butoh master Akira Kasai once told me "it isn't a Dance unless somebody dies. It might take only a second, it might take four hours, it just takes as long as it takes." We are dying to dance and dancing to die.

Refer back to one of the great creative movements of this land. Near the end of the Indian Wars, a Paiute named Wovoka (Jack Wilson) had a vision, of a dance that would save his people and bring peace and plenty. It became known as the Ghost Dance and swept through the decimated tribes of the west. Women and men and children danced in circles and sang until they fell into hallucinatory trances and dropped to the ground from euphoric exhaustion. Ancestors and unborn children fell from the sky.

They painted teepees and ghost shirts with their animated visions and made hundreds of short and repeatable songs. Those who danced it were said to be invincible and unafraid—especially amongst the indomitable Lakota and Sioux tribes of the Great Plains where The Ghost Dance was swiftly banned by the frontier government.

The US army finished off the last of that wave of

resistance at the brutal massacre of Wounded Knee. They drove those they were not able to kill into Canada and the rest onto the diminished treaty lands we know as reservations.

I bring up the Ghost Dance because it is such a clear example of creative resistance. There are many others—ACT UP of course, DADA and the Zapatistas. But the Ghost Dance reveals to us something very special, that very particular web woven by singing and dancing on a stage where the stakes are as high as they could possibly be.

At The Church of Stop Shopping we always look for that high-stakes stage. After 10 years resisting consumerism and the neoliberal economy we turned to that immense and awe-inducing stage of Planet Earth—plundered and pillaged by 7 billion humans and the capitalist imperative, in the midst of an extinction wave of unknown proportions, on the edge of climate catastrophe.

We went to work for the Earth using the most immediate resources we could—our bodies, our voices, our songs, our dances, our visions. And what did we see there, in the exalted state of our creative resistance? We saw that we cannot be radical alone: we rely on the complexity of life, the food chains, the air scrubbing trees, the rains and ancient wells, all those things that remain mysterious. We cannot be radical alone, we rely on skills and dreams and memories that are different from our own. We cannot be radical alone, it has never been more important to work in a diverse community, and amongst a broad range of experience alchemy is everyday. When we are radical together our spirits are sustained by resistance even if resistance is futile.

But no more waiting around for it—we go to the Earth and we ask the Earth, what are you telling me? What do you want me to do? How can I work for you today? And that is how we stay radicalized. We are performers, champions of sublime excitation. It's our job to get carried away, dead from dancing.

The Guyaba Standard

Steve Sprinkel

Guava Paste futures rose today on news that the
World Bank had agreed to reimburse Indonesia and Honduras
for hidden charges in agricultural loans made to
military juntas who landed finders' fees for
indebting rural communities—coerced into growing
export crops requiring high rates of luxury chemicals.

Margarita Rosa Tirado, executive director at the
Bogota Tropicals Exchange, said in an official statement that
"Our foods continue to gain respect with regard to overall
human and fiduciary health. After all, we cannot eat gold,
and guava paste, you must admit, is just as beautiful."

Honey and Cranberry Sauce, closely linked to their
Southern Hemisphere commodity cousin, notched steep
gains as well, as foods continued to outstrip traditional
wealth havens utilized to park exorbitant capital.
Incipient protein indicators in grasshoppers, caterpillars and regional
larvae showed continued promise as the solar and bio sectors (SOBIO)
surged on news that the McDonald's Corporation would introduce
the MacHopper sandwich for select populations.

Gold's free fall continued to drag down the precious metals sector, on
news that the Italian Unione Nacionale of jewelry artisans
unilaterally voted (881-2) to abandon the use of gold and silver and utilize
renewables like nut woods, recycled plastic and bamboo in their place.
"Nobody died when I wore walnut", the viral message spun spontaneously by
international spokesperson Natalie Portman, seemed to sober
South African Mining Minister Keith deGrinder, who responded that
"Once gold lowers to $317 USD per ounce all chemical mining will have to cease.
We will no longer be able to recover our costs."

Indigena International immediately released a statement saying:
"A hundred rivers and all their turtles say thank you to those who have
abandoned the death metal. Our birds will come soon and thank you personally."

Down over 500% from its all-time high of $1806 USD per ounce in 2013,
gold (AU) broke the feared $400 barrier, falling $22 per ounce to $397.
Other vital gainers benefitting from Gold's woes were
Peanut Butter (PBT), Frozen Lima Beans (FLB) and Quinoa (Q).
Spearmint Extract and Basil Pesto remained unchanged as the
North American harvest began. World quinoa producers were
buoyed by the news of steady price support despite a report
finding that world-wide quinoa acreage increased 44% this year.
—Armenia, Colombia

"Corporate feudality has taken the place of chattel slavery and vaunts its power in every state."
—James B. Weaver, *A Call to Action*, 1892

a WORLD of FOOD
by MARJORIE THORP

THE GREAT BULK OF OUR DAILY **COFFEE** COMES FROM BRAZIL, COLOMBIA, CUBA, EL SALVADOR, NICARAGUA and VENEZUELA.

— WHILE COFFEE IS GROWN IN OTHER PARTS OF THE WORLD, THE PRODUCT FROM THE ABOVE NAMED COUNTRIES IS DEEMED <u>SUPERIOR</u>:

A-H-M-M-M

MILK-FED CHICKENS
ARE CHICKENS THAT HAVE BEEN FED FOR TWO WEEKS BEFORE KILLING ON A MIXTURE OF GRAINS AND SKIM-MILK, OR BUTTERMILK. THIS SPECIAL FEEDING PRODUCES WHITER, MORE TENDER FLESH AND GIVES IT A FINE FLAVOR—

The AVERAGE AMERICAN, ACCORDING TO GOVERNMENT FIGURES, CONSUMES **2150 POUNDS** of **FOOD** A YEAR! HIS "BILL OF FARE" INCLUDES

1010 POUNDS DAIRY PRODUCTS
537 " FRUITS & VEGETABLES
193 " MEAT, FISH, EGGS
118 " SWEETS !

Banana Republic

Steve Sprinkel

Seems like yesterday that Maude assigned me a report on Ecuador.
Only 51 years since I sat next to Lance Markwell (he got Argentina),
trying to be as smart, imitating his deft left hand
with my awkward right, dusting off the ancient Colliers,
getting laughed at for saying Kwitoe was the capital
and claiming that they grew sugar cane and bananas
but remembering Thor Heyerdahl and Easter Island
saved things and I got a B.

Bananas are still where they belong, wrapped in blue bags
along the Bula Bula, competing with cacao and the mango
for flat earth near nice water, east of Guayaquil.
Working west you go to school on deforestation,
as clouds fly east from the generous ocean
over bramble, red dirt, cactus, and wind blown vanities
until the river solves the question.
In the free hills of Manabi, home to
the latest smoke-choked land rush for the bananas
of the moment, jabbed in the earth while the big trees smolder,
we repent of an undoing too dreadful to remember.

Would it really help to haul 15,000 twenty-five horsepower
electric wood chippers running on direct source 25 Kw solar arrays
to the city of Pedro Carbo, Manabi Province,
and trade all that global smoke for so much God-awful racket
the birds all die because they can no longer
hear each other well enough to mate?

One thing I can discern now, Maude, still tentatively,
is that not too much has really changed in these five decades
despite the sobbing and salvage
so a lot less needs to be unwound.
This blue haze is nothing compared to the
big doom hanging close to Houston or Guangzhou,
where Romance and Science deserve a better test.
—Pedro Carbon, Ecuador

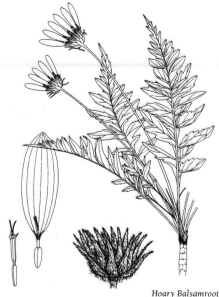

Hoary Balsamroot

Plant The Seeds Of Memory, Or, Why Archive?

Rick Prelinger

Every day we break new ground that others have seeded before. This Almanac testifies to what we have learned from millions of yeoman farmers, scientists, hackers and activists who have lived and worked on the land before us. But history, like land, requires stewardship. Knowledge needs cultivation to pass from generation to generation. Memory of our achievements, inventions and struggles will endure only if we make an effort to remember and record.

From Activist Archivists of New York, here are five reasons why we need to collect, preserve and propagate the history of our lives and works:

WHY ARCHIVE?

Accountability — Archives collect evidence that can hold those in power accountable

Accessibility — Archives make the rich record of our movements accessible. We can use them to ensure transparency, generate discussion, and enable direct action.

Self-Determination — We define our own movements. We need to create and maintain our own hisorical record.

Education — Today's videos, flyers, web-pages, and signs are material for tomorrow's skill-shares, classes, and mobilizations.

Continuity — Just as past movements inspire us, new activists will learn from the experience we document.

EDITOR'S NOTE Greenhorns is increasing our own efforts to collect and preserve the archival record of young farmers and the new agrarian movement, starting with a collaboration with Rick Prelinger who is collecting archival footage of rural landscapes to put together for a LOST LANDSCAPES of Rural America. If you have home movie footage that you would like to have included in the project then get in touch with him through the Prelinger Library website, at www.prelingerlibrary.org

FACTS AND FACTORS IN EUROPEAN AND AMERICAN LAND TENURE

THE INHERITANCE OF FARM LAND

INHERITANCE laws, according to Toqueville, ought to be placed at the head of all political institutions.[1] In European countries land inheritance is regarded as the heart of the tenure system. According to Georges Sorel, "the inheritance of land is so closely connected with the juridical sentiments of a people that it is possible to use different inheritance laws as characterizations of different societies. It is there that the local laws fight their last battle against the unifying tendencies of centralized governments, and sometimes, win them."[2] In the United States the importance of land inheritance has not yet reached its acme. It is growing at a rate in proportion to the increase of the time distance between the first and the present generation of settlers. In most sections of the country this period exceeds another generation's lifetime.

Inheritance and the Distribution of Wealth.—It is no exaggeration to say that the inheritance customs which prevail in a region determine the social structure of that region to a large extent. Inherited wealth, said Hobhouse, is "the main determining factor in the social and economic structure of our time."[3] Inheritance in all its forms is one of the most powerful factors accounting for the unequal distribution of wealth. As H. D. Dickinson has shown, this cause of the unequal distribution of wealth operates with a much higher intensity than the others and is cumulative in a way the others are not.

Incomes derived from work are increased with progressive difficulty, whereas those derived from property are increased with regressive difficulty. . . . While one who relies on personal exertions for his livelihood finds, after a certain point, increasing difficulty in increasing his output of efficiency-units, one whose income is due to legal claims is able, owing to the diminishing marginal utility of income, to increase the number of those claims the more easily the more he already has.[4]

[1] *Democracy in America*, I, 49.
[2] *Introduction à l'économie moderne* (2e éd. Paris, Rivière, 1922), p. 74.
[3] *Liberalism*, p. 197.
[4] *Institutional Revenue* (London, Williams & Norgate, 1932), pp. 173-74.

COOLING AND SHIPPING MILK

Grange Activities

OLIVER D. SCHOCK, PENNSYLVANIA

The Keystone grange exchange, an organization of members of the state grange, was chartered at the state department at Harrisburg, with a capital of $10,000 to do business in Pennsylvania. The central office will be located in Harrisburg. The object is to bring the producers, consumers and railroads into closer business relationship. The $10,000 is divided into 5000 shares at $2 each, and state grange leaders subscribed for the entire amount.

A considerable number of the scouts and field agents employed by the Pennsylvania chestnut tree blight commission are college students, who will shortly resign to resume their studies. Many of these are forestry students.

There is every indication that fall fairs in Pennsylvania will be unusually well patronized. Fake and immoral shows, as well as gambling outfits, have been generally eliminated, and a higher moral tone and educational features have added to their genuine value.

The granges of Delaware and Chester counties held their annual reunion and picnic at Lenape park, on the banks of the Brandywine, last week. Fully 6000 people were present, and a day of pleasure and profit followed. Prof Hunt of the state college was the orator of the day.

INITIATIVE AND REFERENDUM.

When by direct legislation the voter has a chance to deal with measures as well as with men, there will be a new and purifying interest in politics. Under our present system the candidate is all important, says the Nebraska Independent. If the railways or great corporations control the conventions of all parties and obtain candidates who will be their servants, the voters are disgusted and become indifferent, for they realize that the election can not evolve good out of the evil that began with the conventions or even at the primaries.

Under the initiative and referendum the legislator will not be as important as he is today. If the legislature fails to adopt a law favored by a certain proportion of the voters, they can have the measure submitted to a vote of all the people. If the legislature passes a law displeasing to a large number of the voters, they have the same recourse, for no law passed by the legislature would go into effect until a reasonable time had elapsed—say thirty days for a city, sixty days for a state, and four proposition, by signing and filing a petition, and if the legislature proves hostile to the proposition then it goes to a vote of the people, a majority enacting or rejecting.

The indifferent voter is a serious problem of the times. Many good men have abandoned politics in disgust. Under the initiative and referendum their interest would be rekindled. When they understood that their votes would have direct influence for or against certain measures of legislation, they would become devoted to politics and therefore to the public welfare.

The educational value of the system would be inestimable. A minority could force the general public to consider some measure which might seem the hobby of a school or faction. If the measure proved too radical it would be defeated, but under any circumstances the voters would derive great benefit from a real "campaign of education."

The majority would still rule, but the minority would not be submerged. While true democracy requires that the majority shall rule, there is always grave danger in giving the minority too little power. In this respect the initiative and referendum would have a correct influence. A minority is frequently the saving element in society, and a system which permits the minority to make itself heard can not but improve political conditions in any democracy. In the long run, of course, a democracy must place its faith in the wisdom of the majority, and it will go well with that nation which adopts the system best fitted to keep the majority enlightened. "We need never fear an error which reason is free to combat," said Jefferson, and the American people should not fear a system which will give reason a free field and error no favor.

September In The Field

Douglass DeCandia

The Sun Rose painted
Love in lines of juvenile light
onto the soft darkness of dawn.
Shadows showed themselves slowly,
shyly, holding tight
to the cold of an Autumn night;
and the cold held me
sweet by the hand
like an old and familiar friend.
At the kitchen table,
tea leaves steep in hot water
as the thin lines turn to day
and the cold, smiling, says "so long",
soon to return, soon to stay.

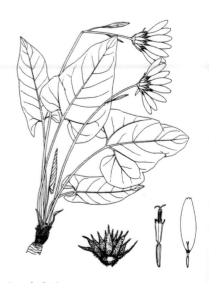

Arrowleaf Balsamroot

Into The Growing Shadows

Douglass DeCandia

Autumn is introduced
again, like an old friend
by the cool morning air of early August.

It has brought along a suitcase
filled with magic and with memories
that it will unpack slowly over the next few months;

colors it soon will throw upon the forest
inspiring the artists by humbling them,
sounds by the darkness choir
seduce the ears of trained musicians,
smells that bring the old and faded back to youth.

In the fields, Autumn shows itself
in the greatest collaboration and
communion between man and nature;

the single seed we planted in Spring,
Summer has grown to a plant,
bearing the fruit that Autumn now cures, sweetens,
holding in each a thousand seeds,
a thousand plants tomorrow.

How can we be hopeless
when there so much hope found in one tiny seed?
Autumn,
like an old friend,
reminds us
what is possible.

Subsidies Play Big Role In U. S. History

Government subsidization of industrial, agricultural, and consumer interests has played a prominent role in American economic life since the foundation of the republic. The very first Congress, in 1789, passed a law in aid of the American merchant marine through the remission of a certain percentage of the customse duties llevied on merchandise if c a r r i e d in American ships. Not many years afterward (1796) the long-standing p o l i c y of aid to transportation began/ when a private citizen received a grant of land as a means of aiding the construction of a highway. In the early years of the nineteenth century, Congress made extensive grants of land and money for the construction of roads and canals. The famous Cumberland road was financed in this way, as well as the Louisville and Portland, Chesapeake and Ohio, Chesapeake and Delaware, and Dismal Swamp canals.

The railroads, especially the western railroads, have been large beneficiaries of government subsidy. By June 30, 1941, some 179,284,978 acres of land had been donated to railways in aid of construction. Of this total, 130,401,606 acres came from the Federal government, and the rest from the states, which, however, had received most of this land from the Federal government originally. This represents public aid to an amount estimated to be approximately $492,000,000. In addition to these land-grants, land valued at $87,000,000 was also given to the railroads for rights-of-way.

The policy of subsidizing shipping was continued throughout most of the nineteenth century. After 1847, it took the form not only of tariff remissions but also of direct subsides to the mail. Between 1847 and 1877, Federal aid to ocean mail ran to some $21,000,000. Between 1920 and 1940, around $750,000,000 were spent through various agencies to subsidize the merchant marine.

Federal expenditures in aid of domestic water transportation had cumulated to $2,500,000,000 by June 1940. About half of this amount was disbursed in the decade 1930-1940.

Air transportation had also been the recipient of extensive government aid. In the period 1931-1941 inclusive at least $100,000,000 had been given to the industry through mail contracts, and at least $270,000,000 had been expended by 1941 on the construction of the airways system and of airports.

The government has always pursued a policy of disposing of its public land rapidly and cheaply, which has in effect constituted a subsidy to agriculture. In 1862 a policy of outright gift of public land was inaugurated under the terms of the Homestead Act. Over 246,-000,000 acres, or almost 18 per cent of the area of the United States, has been given to homesteaders. Land grants, as well as appropriations of money, have provided for the establishment of agricultural colleges and agricultural experiment stations.

Farmers

OPA's price reduction of approximately 10% to the consumer on butter and meat is now being paid to the processor who in turn reimburses the farmer. If these reductions were rolled back to the cattle and hog raiser, and the price of hogs and cattle were reduced accordingly the supply of meat and butter would be retarded at a time when the maximum supply is needed. 450 million dollars has been approved for these subsides.

At the present time the Commodity Credit Corporation has allocated 60 million dollars to be paid direct to dairy producers who produce and sell milk during the October, November and December quarter of the year.

Through the depression years, the Agriculture Adjustment Administration got farmers to increase their soil conserving crops and practices. Now their lands are in better productive condition for all-out production for war. Farmers benefitted by subsidies being paid to them through the soil conservation and parity programs. They were encouraged to increase soil productivity by carrying out soil conserving and water conserving practices.

Metals

Since metals are so essential to the war programs, the Metals Reserve Corporation is encouraging the production of various metals in this country. They are buying them at a high price and selling at ceiling price-abnormal freight rates at cost—approximately 25 million dollars—will be paid on metals imported from Latin-America and abroad this year—this is a form of subsidy. Manufacture of synthetic and purchase of natral rubber is being subsidized. Metal Reserves has been paying premiums that will cost approximately 53 million dollars a year on certain excess domestic production of copper, lead and zinc since February 1, 1942. And due to efforts to further stimulate production, this cost may increase to as much as 80 million for 1943.

The government has paid approximately 5 million dollars to manufacturers to obtain the necessary production of aluminum rivets used in making air-planes. This was paid to manufacturers at a high cost because they are not normally in the business.

Petroleum

Petroleum products for 1943 will total approximately 95 to 100 million dollars in subsidies.

Due to submarine warfare, transportation costs on coal to New York and New England will cost 25 million in subsubsidies.

As early as 1891 ocean-going mail was subsidized to approximately 20 million dollars. Second class mail privileges to newspapers has always cost the postal department more than was received.

October

IT TAKES
A VILLAGE
TO RAISE A FARM

Spirit & Dominion

Chapter 10

Farming The Ecofeminist Life

Alison Parker

I blame it on Laura Ingalls Wilder—I always wanted to be a pioneer. If and when my family did have a yard, it was always small. I remember perching by the window, reading about Laura and her sister Mary harvesting potatoes and turnips for the winter. I looked out the window at the long sidewalks and rows of townhouses, and fantasized about planting potatoes and turnips in the flower boxes below.

So when I first heard someone talk about growing food as an act of social and environmental justice, it was like a lightswitch clicked on. I was attending Earth First! meetings as a freshman in College, but it had never occurred to me that growing food could be a radical and political act.

The what's-for-dinner question in my family is not just between Thai take-out or pizza, it's about the food's literal and figurative roots. Were these beets produced in healthy, chemical-free soil, unattached from ghosts of thousands of truck miles? Did this chicken live a wandering life eating bugs, raised by a farmer we know? Did this honey come from our hives, or those of a holistic beekeeper nearby?

My senior project at Evergreen State College was on Ecofeminism. A key ecofeminist tenet is that the use of agribusiness chemicals is a feminist issue. Women are still the primary growers of most of the world's food, and women are the ones dealing with deadly poisons like DDT—still legal in places like China. If we, as feminists, are buying cucumbers with exploitive, toxic origins, we are taking part in the system we are working so hard to unthread. If you are eating a pulled pork sandwich at a local deli, the pork has often come from a suffering animal in a devastating industrial factory farm. That tomato in your run-of-the-mill café salad had a harrowing journey from Mexico, its poison-sprayed body wreaking havoc on the lives of farmworkers and the landscape.

Ecofeminism asks us to take a step back and support non-exploitive, non-destructive enterprises. I can do this by growing organic food for my family and community. It's important for me to move away from capitalist economics and their corporate greed, climate change, and environmental destruction. As feminists, these are the oppressions we are up against. Canning tomatoes may seem trivial, but it's one less poison tomato grown.

This "New Domesticity" has its critics. We can't forget to look at these movements through a race/class lens. We can't forget that CSAs and organic farmers' markets

our "radical homemaking" way of life sound miserable. This is fair: farming is no picnic. Washing carrots outside in December is painful. Canning forty jars of pickled beets takes forever. But for me, it is rewarding. I save money, and my values remain intact. I am not succumbing to consumer-driven cultural expectations that I oppose. I view wealth as something besides dollar bills; we barter, we grow, we forage, we hunt, we dumpster dive, we work-trade, and we never feel empty or dissatisfied for lack of purchasing.

But this isn't the only way to be part of the ecofeminist revolution. It works just as well to buy and barter from people and places you want to support. At our CSA people can work in exchange for their vegetables, and we are not the only farm that does this. Shop at the farmer's market, buy raw, fermented sauerkraut directly from a farmer, knit a sweatshop-free hat, ride your fossil fuel-free bicycle, cook something simple from scratch. These are feminist acts. They may be small but they oppose and un-weave heartless systems of oppression. These oppressive systems are the prison walls, not those of your kitchen.

are still more prevalent in affluent areas. Here in the Midwest CSAs are relatively new, and take time to move elsewhere. In Chicago, Growing Homes and other non-profit CSAs bring their services specifically to food deserts. My farm and many others will accept food stamps. Vacant lots can be blank canvasses for food growing. I have hope that like an inkblot, these things will spread in all directions.

But there is debate. In the Washington Post Emily Matchar embraced the virtues of canning and the domestic arts as a young woman, and was met with disgust by writer Jamie Stiehm. Her counter-article bemoans the step backwards for women that the New Domestics bring, and grumbles "there's no need to make a fetish out of all that. We must pursue progress for women, given all we have been given." But Steihm is actually the one taking a step back. She dismisses women's choices along with an entire traditional and political context.

My husband and I now own and run Radical Root Farm, an organic CSA and market farm outside of Chicago. To some, many aspects of

POSITION OF EARTH, SUMMER AND WINTER.

THE SCHOOL HOUSE

Every district school house should be a community meeting place. Until the Farmers Union realizes its dream of strong local with its own hall in every township, the school house is practically the only building that is available as a social center in many parts of the state.

Not a week should be allowed to go by without some sort of gathering to promote the better acquaintance of the people of the neighborhood with each other. There are scores of questions of timely interest that should be fully and freely discussed, by all Kansas farmers during the coming winter. The Farmers Union local will do its share in the educational work but auxiliary to the local there should be a debating club and the men and women in the neighborhood, old and young, should be encouraged to take part in the debates and discussions.

Occasionally during the winter a speaker from outside the immediate neighborhood should be asked to lecture upon some subject that he is known to be a master. The expense of such a lecture need not be so large that it cannot be met by a small admission fee or by some sort of social or supper. It pays to talk things over with each other and it also pays to get the views of an outsider once in a while. In this age of the world no one can afford to live in a shell.

WOMAN'S FARM WORK.

Many a farm and farmer would go to rack and ruin, morally and financially, were it not for the woman's influence and the wife's help. The average farmer and farm are what they are largely because of the wife.

Sound health is fundamental in human success. Material and ethical development depend upon this, for while individuals in poor health and with feeble digestion, may accomplish much and enjoy life, we all know that for the mass the vital essential is an organism sound and normal from head to foot.

To properly nourish the body and wisely to insure its health, physical and moral, are duties in every family that devolve upon the wife. To fulfill her duties in this respect, a woman requires genius of the highest order, intuition, knowledge, experience, tact, sense, patience, all the attirbutes of perfect and successful motherhood and womanhood. Few men on the farm understand the qualities possessed by the women who toil by their sides. Every woman possesses these qualities to a degree; life at its best depends upon the extent to which she is able to develop and employ them.

The responsibilities of women are complex and they must fit themselves to do their work. The educational needs of young women who are to do the work of the farm are but feebly realized and poorly provided for in schools. The manner in which a woman rises to her responsibilities on the farm, in view of her inefficient preparation for her duties, is phenomenal. The successful woman on the farm has the most glorious success possible in a woman's life. Woman's work on the farm is just as important, just as satisfying, as man's.

home

Fungi

Sci - Fi Farming

Tess Brown-Lavoie

Farming in America is a science fiction project. We re-imagine a future outside of current cultural norms established under capitalism—where the tenets of consumer culture are valued far above labor, ethics, environment, or community.

Capitalism, patriarchy, whiteness—call it what you want—our dominant politics are oppressive. Agrarian life is not outside of capitalism, but it fundamentally challenges mainstream culture. Food production demands a commitment to place that is not demanded of practitioners of most industries. Amendments—

feed, fertilizer, good weather—create growth on farms. Contrary to the liquidation ethos of big business, wealth does not materialize from nothing. Farming incorporates ethereal terms into its logical formulas—the value of products, work, sweat equity, and goodwill. Somehow, these pay the bills.

In my favorite sci-fi books, injustice and amorality are revealed by stark depictions of the fictional world. I can explore my own relationship with the environment through water conservation on Arrakis in *Dune*. Shevek's appreciation for Urrasti trees in *The Dispossessed* teaches

more than newspaper articles. Similar revelations occur at my farm. Sun on my arms, water in my throat, work upon my body are clear sensations, and I am struck by my good fortune—to live where there is still rain and sun, heat and cold, where generosity, stewardship, work, and food are valued, where the body is a tool. When I bike back into the city after dealing in agricultural currency all day—spending manure, favors, and care in exchange for pounds of fruits, soil, greens, and roots— the border crossed is palpable. The farmers market is my currency exchange—vegetables into cash. Cash earns admittance to bars and restaurants, and pays rent on my apartment.

Farms sell food, but farmers know that capitalism is only marginally helpful in understanding work on human-scale farms. These farms subvert capitalist norms in several ways.

Generosity

Generosity, an unquantifiable economy, is a quiet motor in agricultural communities. Agricultural mentorship has a long genealogy, and favors form decades-long chains of reciprocity—tools and talents shared in farming communities. Generosity cannot be commodified, yet it is a powerful engine of community.

Energy

Energy is bought and consumed on fams, but also produced, and reproduced in the form of compost, pasture, milk, apples, and short ribs. Wendell Berry's "Uses of Energy" offers a brilliant discourse on this idea. Healthy farms have closed energy cycles, operating outside of capitalist economies. A tank of fuel is instant gratification, compost is nutrients bought on layaway.

Economics

Economics and standard market forces—rules of scarcity, surplus, and demand—all affect prices for farms. But in integrated community agricultural businesses diversity mitigates risk, and success of the farm is not totally dictated by market forces. Weather, disease, accident, and other natural phenomena all influence agricultural production. The logic of our world is defined by our economic paradigm, but this logic cannot instruct sustainable farming. The bottom line is only one thread in a fabric of agricultural decisions. There are other forces at play—enough so that science fiction might better interpret these farms.

Like science fiction, farming engages with the potential of the future. To farm is to invest in future productivity—a seed is worth much less than a watermelon. Farmers know the decisions made in spring dictate fall yields— instead of investing money we don't have, we invest ourselves in the authorship of a more pleasant (or feasible) future. We use our bodies, our land, our rich soil, our jars of tomatoes to build a landscape of fruit-laden vines, healthy animals, and soup all winter.

We must be active and intentional authors. The particulars of our stewardship should manifest the changes we wish. I want a utopia with beauty all year, where waste, 18-wheelers and winter anemia are memories. I want a sisterhood of knowledge production and dissemination rooted in passion and necessity. Farms are stages upon which futures are tested, and I want open-source results. Resilience is key, innovation is necessary, sustainability is given. Community is precious, because living in a sci-fi world can be a solitary project. I have visited many utopias. I treasure these experiences—they guide me in my own agrarian authorship.

Sci-fi reimagines the normal. As a female-identified farmer, I am invested in the corruption of the critical but limited role of "farm-wife," historically in charge of canning, care-taking, and keeping the books. Ownership expands a woman's role on a farm to include physical strength and intellect. Ownership enfranchises—it will be a benefit our communities to include more people in this. The farms I visit are often operated by white people. But urban agricultural utopias are more diverse, proving that when agriculture is accessible the farm benefits from a broad buy-in. Diverse people bring diverse practices and personalities. Accessibility—both to good food, and the opportunity to grow it—is a concern of this world, and this moment. The answer must address the rules of hierarchy and privilege. May the next chapter grow in the fields where rules are actively resisted—on science fiction farms.

THE ATMOSPHERE ONE CARRIES.

Nature's forces carry their atmosphere. The sun gushes forth light unquenchable; coals throw off heat; violets are larger in influence than bulk; pomegranates and spices crowd the house with sweet odors. Man also has his atmosphere. He is a force-bearer and a force-producer. He journeys forward, exhaling influences. Thinking of the evil emanating from a bad man, Bunyan made Apollyon's nostrils emit flames. Edward Everett insists that Daniel Webster's eyes, during his greatest speeches, literally emitted sparks. If light is in man, he shines; if darkness rules, he shades; if his heart glows with love, he warms; if frozen with selfishness, he chills; if corrupt, he poisons; if purehearted, he cleanses. The soul, like the sun, has its atmosphere, and is over against its fellows for light, warmth and transformation. This mysterious bundle of forces called man, moving through society, exhaling blessings, or blithings, gets its meaning from the capacity of others to receive its influences. Standing at the center of the universe, a thousand forces come rushing in to report themselves to the sensitive soul-center. There is a nerve in man that runs out to every room and realm in the universe. Man dwells in a glass dome; to him the world lies open on every side. Each man stands at the center of the great network of voluntary influence for good. Rivers, winds, forces of fire and steam are impotent compared to those energies of mind and heart that make men equal to transforming whole communities and even nations.—N. D. Hitlis.

WHAT IS PARITY?

The farmer, in order to honestly weigh the merits of various legislation which would forbid fixing of price ceilings on farm products at less than full parity, or the dangers of legislation which would include the cost of farm labor in computing parity, must first of all understand how parity works.

Parity is simply an exchange formula with manufactured goods applied to farm prices.

Parity seeks to establish a fair value for the things the farmer sell as against the things he must buy.

It is as though the commodities the farmer sells were put in one pan of a huge balance scale and the articles he buys were placed in the other pan. Parity keeps the scale balanced. Prices may fluctuate, but they fluctuate evenly, so that a balance is maintained.

We must remember this, that parity is a sound formula. The farmer's price troubles have not been due to parity or any failure of the formula, but have been due to the inability of the farmer to achieve parity for most of his farm products until recently.

To reach a fair value for both the commodities the farmer sells and the items he buys, the Congress selected a period 1909 to 1914 as the base period. This was when the prices the farmers received for their commodities were considered fair in relation to the articles they bought.

To establish a balance, the Congress drew up a list of 86 articles the farm family uses, added another 88 items that are used by farmers in prod-uction, and added interest and taxes. Next it decided how much farm products, such as wheat and corn, and beef and milk, was required to buy all these 174 items and meet taxes and interest, in this base period from 1909 to 1914.

For example, let us say that between 1909 and 1914, the farmer had to sell 100 bushels of wheat to buy a half-dozen items to run his farm. If wheat were selling at parity today, he could still buy the same half-dozen items for 100 bushels of wheat.

The parity formula would be adjusted so that the price of wheat, as one of the basic commodities raised by farmers, would rise and fall just as much as the 174 items which are used to compute parity.

Keep in mind that parity is the goal of American agriculture. It ignores how much prices rise and fall, and tries to stabilize the exchange-value of farm products. It makes the things the farmer grows the real medium of exchange, rather than gold and silver and bank notes.

It may require occasional adjustments. But these should be wisely made, and for long-range benefits, not immediate and shortlived gains.

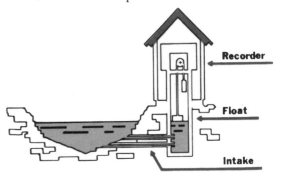

Recorder

Float

Intake

Regrets: When Your Past Stays With You

Shannon Hayes

I was meeting with the school psychologist over my daughter's vision therapy program last week when a wail came from down the hallway, and grew progressively louder and more despondent. Occasionally the words "I WANT MY MOMMY!" could be separated out from the cries of despair. My mother's heart wanted to jump up, run to the child, and throw my arms around her. Her tears were contagious. I felt my own eyes growing wet.

The psychologist paused in her conversation. "That sounds like one of the older kids," she said matter-of-factly. We did our best to resume our meeting while the lamenting outside the door ensued.

A moment later, the school principal popped her head in the room. "I'm sorry to interrupt you," she spoke softly to the psychologist, "But one of our students made a bad choice this morning, and I think she could use a little of your time."

I like the way she worded it. "Made a bad choice." In our family homeschool environment, I am often out of the loop on the professional language used to describe the actions and regrets of children. Our expressions for bad choices are more colorful, in keeping with farm vernacular, and nowhere near as gentle. She made it sound like, whatever happened, it wasn't really so bad— just a little slip up that would soon be forgotten. But from the caterwauling I heard in the next room, I don't think that child's bad choice would soon be forgotten. In

a rural community like ours, a bad choice is often glued into the memory book, a page in the album of everyone's recollections that will follow that child as long as she lives here. At forty, I still meet my teachers, the parents of my friends, and my fellow school mates in the library, at the coffee shop, at the bank, on the sidewalks in town, at local concerts. Many of them have managed to remember (and remind) me of my earlier choices long after I've forgotten them myself.

I am certain that child was crying over the horror of the moment she just experienced. But I am certain that she, too, recognizes the permanence of it in the memories of her classmates and teachers. That, I am guessing, is where her deeper pain lies.

Someday, after she graduates, she will make an important decision. She will be free to leave every bad choice behind her, find a new place to live, and grow into a new identity, liberated from the humiliations of her past.

Or, she will choose to stay, and let her bad choices become a thread in the fabric of her identity. I think about my own decision to stay here, in the same community where I grew up. Today, I see myself as a strong woman, capable of making good decisions and taking care of myself and my family. But in spite of my positive self image, all around me are people who have watched me grow, who have known me during weaker moments, when I failed to take care of myself, when my decisions were faulty. And I know that it is easy, in a small community where everyone knows each other, to allow that past to define my identity.

How do we choose to stay in one place the entirety of our lives, knowing that the people around us have all borne witness to our follies and imperfections? How do we develop into our true selves when we are surrounded by people who think they know us better? How do we become who we want to be in a place where there is no such thing as a fresh start?

The secret to growing into ourselves with no fresh start, I think, lies in two things. First, our need for attachment supersedes our bruised egos. Belonging to a place, to a group of people, is too important to allow our ties to be fractured by bad choices. The second secret is the reciprocal knowledge. It is true that my family, friends and neighbors have a clear picture of my colorful past. But I know many of their own secrets. These are not scurrilous defenses against blackmail. Rather, they are banks of knowledge that we carry by virtue of long-standing connections. In many cases, we know each other for a lifetime. And that helps us to recognize how ubiquitous bad choices are in everyone's life path. More importantly, when we pair the drive for human attachment with the long standing experiential knowledge that nobody is perfect, it becomes easier to forgive ourselves. And the ability to forgive ourselves is the key to moving forward in one place, with no need for a fresh start.

Wending
Qayyum Johnson

The rain finds itself in the air
discovers the branch
immolates & dispenses,
upon the grass beaded,
earth the darker
for its lusty drink.
Tipple your fiddle old man
& get all the young people dancing
we need to scratch the new floors,
have the hall ring with heels
voices, bodies, music.
Let's together make joy
enough for everyone
to go home soaked tonight.

The Criminalization Of Seed: Notes From Columbia

Kat Shiffler & Erica Rojas

In 2011, the Colombian government authorities ransacked the warehouses and trucks of rice farmers in Campoalegre, in the province of Hulia, and destroyed 70 tons of rice. It said the rice was not processed according to the law.

This law, known as resolution 970 and drafted under Free Trade Agreements with the United States and Europe, created a legal monopoly over rights to seed sold by private transnational corporations from these countries. This meant that farmers caught selling farm-saved seeds or simply "unregistered" native seeds could face fines or even jail time.

The massive public outcry—the National Grassroots and Agrarian Strike—began in real earnest on August 19, 2013. Over 200,000 potato, rice, fruit, coffee, dairy and livestock farmers; miners; teachers; truck-drivers; healthcare workers; and students went on strike and blocked roadways in 30 key corridors around the country.

Citrus farmers in Valle del Cauca dumped 5,000 tons of oranges onto the highway and dairy farmers in Boyacá poured out over 6,000 liters of milk. Farmers in the countryside stopped producing food for the cities of Colombia in a massive demonstration of the frustration of the small-scale farming sector at the neoliberal policies that are pushing them out of existence.

Their demands included suspension and renegotiation of the U.S.-Colombia FTA and support for small-scale agricultural production, as well as access to land; recognition of campesino, indigenous and Afro-descendant territories; the guarantees of political rights of rural communities; and investment in rural areas.

As potato farmer Cesár Pachón said, "We're not asking for more money. We're asking for conditions and agricultural policies that allow us to survive."[1]

INVOLVED PUBLIC

After 21 days of a farmer-driven massive general outcry, the government was forced to negotiate and eventually suspend Resolution 970—the seed law—for a period of two years. The suspension will only apply to domestically-produced seeds, not imports and the government is now supposedly writing a new small-farmer friendly law.[2] Colombian civil society continues to reiterate calls for the Resolution to be completely repealed.

Four people were killed in the upheaval, and nearly 200 injured.[3]

---

1   Duranti, Julia "A struggle for survival in Colombia's countryside" http://www.bilaterals.org 30 Aug. 2013
2   GRAIN "Colombia farmers' uprising puts the spotlight on seeds" Sept. 2013
3   According to data from newspaper, El Tiempo.

The following are some thoughts from Erica Rojas, a campesina from Boyacá. She is also a sociologist and current doctoral student in Agroecology at la Universidad de Antioquia.

It is true that farmers in Colombia are naïve. We endure and we are noble to the limit. But it is worth remembering that the last time these naïve Colombian peasants rebelled, they achieved the independence of several Latin American countries in what was called the Battle of Boyacá.

Today, two centuries after we acquired independence from Spain, we realize that we are yet again dependent on foreign powers. This time it is the economic policy of the United States—the country who determines the free trade agreements that make it a crime for peasants to retain their own seed. This leads to the destruction of traditional production systems and the disappearance of the peasantry.

Currently the negotiation process between the government and farmers through the peasant movement has entered a period of crisis and stagnation. We do not have the institutional and state dynamics to meet the expectations and demands of the peasants. We talk on the radio, TV and social networks to restart mobilizations in the absence of an institutional or legal option, but we don't see a scenario of long-term solutions.

It is important to mention that since the start of talks between the government and farmers, the government has signed new free trade agreements with Israel. The plot thickens further when the U.S. government says that Free Trade Agreements are not modifiable, and that trade relations with Colombia are dependent on large numbers of weapons sold to Colombia and maintaining the armed conflict in our country. Added to this, in the last year Colombia's congress approved a series of measures that criminalize social protest.

Storing and cultivating native seeds has been criminalized not only in Colombia, but also in Mexico, Chile, Argentina, and India. In Colombia, however, not only is native seed conservation criminalized, but the state is using force to terrorize villagers for protesting this offense.

It is also important to note that there are still requests and demands that were left out of the negotiation process, such as the fight for food sovereignty and non-dependence on pesticides, and healthy food production. The peasants do not need to compete to poison the people. The real struggle is to preserve traditional production systems and continue to provide the people with healthy food.

Water Leak

Steve Sprinkel

Water leak tames the swallows,
desperate for mud in this drought.
They load their beaks
at my feet,
slanting in to alight in a flutter,
cutting back out to hit a quick daub,
back again in rare ignorance of me.

I couple hoses,
monitor the long lines in the potatoes,
pushing water to the very end
where a hard swale in the dirt
indicates I could open the line
and make a shallow muddy pond for them again.

There I watched the water arc upwards over
the thick tire tread I struck there when last it rained.

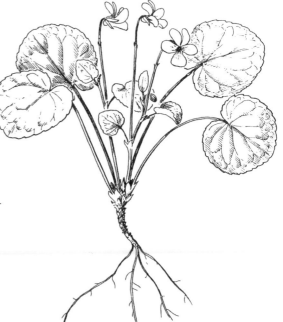

Round-Leaved Violet

A New Society In The Shell Of The Old:
Peter Maurin's Green Revolution
And The Catholic Worker Movement

Joseph Wolyniak

On December 9, 1932, an itinerant French philosopher showed up on the doorstep of a leftist journalist with an idea for a newspaper to propagate the ideals of agrarianism, personalism, communitarianism, anarchism, distributism, and pacifism as the first step towards "building a new society in the shell of the old." The journalist had, just a day earlier, offered what she would later recall as "a prayer which came with tears and anguish, that some way would open up for me to use what talents I possessed for my fellow workers, for the poor."[1] She was crazy enough to entertain the possibility that the disheveled stranger who subsequently materialized on her stoop might literally be an answer to prayer. His name was Aristode Pierre Maurin, though he would be known to posterity as Peter Maurin; her name was Dorothy Day. Together they would publish the first Catholic Worker newspaper on May 1, 1933—which sold for a penny a copy—and thereby formally began the Catholic Worker Movement.

Maurin understood his vision to be the makings of a "green revolution"[2] (not to be confused with Norman Borlaug's Green Revolution), which included the founding farming communes—what he termed "agronomic universities"—wherein "the scholars could become workers, and workers scholars; where a philosophy of work would be restored to people; where [the people] would regain a sacramental attitude toward life, property, and people in relation to them."[3] The idea was that individuals and families would live together in intentional community and voluntary poverty, work the land as a means of provision for any and all in need, and remain committed to the study of the ideas that had brought them there in the first place (what Maurin called "round-table discussions for the clarification of thought").[4] The farming communes were also meant as a means of desisting from the increasingly urban, industrial, stratified, capitalist culture; or as Day put it, "the combating of the servile state."[5] They were meant to provide work to those who might not otherwise have recourse to gainful occupations (and the accompanying dignity it fosters), to reconnect and rehabituate women and men who had become estranged from the land on which they depended, and to provide respite and

---

1 Dorothy Day, *The Long Loneliness* (New York: HarperCollins, 1996), pp. 165-166.
2 Dorothy Day, "Peter's Program." The Catholic Worker, May 1955, p. 2. See also: Anthony Novitsky, "Peter Maurin's Green Revolution: The Radical Implications of Reactionary Social Catholicism." *The Review of Politics*, Vol. 37, No. 1 (Jan., 1975), pp. 83-103.

3 Dorothy Day, "Peter Maurin Farm." *The Catholic Worker*, Oct/Nov 1979, pp. 1, 2, 7.
4 Dorothy Day, *On Pilgrimage* (Grand Rapids, MI: Wm. B. Eerdmans, 1999), p. 149
5 Dorothy Day, "Letter To Our Readers at the Beginning of Our Fifteenth Year." *The Catholic Worker*, May 1947, pp. 1, 3

hospitality to those seeking to escape the grit, grime, and grind of the city.

In successive iterations of The Catholic Worker newspaper, Maurin laid out the contours of his vision, most often in the form of what he called his "Easy Essays"—digestible, even lyrical singsong statements of his philosophy (which he reportedly recited ad nauseam). The idyllic way of life was put not just into print but also into practice, with an ever-increasing readership founding communities that instantiated the vision in sum or in part.

In cities, Catholic Workers founded "houses of hospitality" to provide food, shelter, and clothing to the poor (who were considered not 'cases' or 'clients' but 'guests' and 'friends'); in rural areas, they founded agronomic communes. St. Joseph House and Maryhouse in New York City constituted the first urban

ventures, with Maryfarm founded shortly thereafter in Easton, PA. The movement has since grown to over two hundred Catholic Worker communities throughout the world, with the original communities still extant and still distributing The Catholic Worker for a penny a copy. In addition to their commitment to hospitality, farming, and intentional community, Catholic Workers are committed to various forms of collective action and nonviolent resistance. They generally eschew federal tax-exempt status, seeing it as a co-opting, constraining, and corrupting connection to the state. Each community is autonomous and self-determining with, as Jim Forest notes, "no board of directors, no sponsor, no system of governance, no endowment, no pay checks, no pension plans," and since Day's death in 1980, "no central leader."[6] In an attempt to codify their commitments, The Catholic Worker newspaper drafted a statement of "Aims & Means" in 1987, which most—but not all—Catholic Worker communities have adopted and adapted. A snippet from that statement is included below.

Catholic Workers share commitments that are concomitant with, if not completely indistinguishable from, the various green and agrarian movements of today. They seek to answer the vexing questions of the present with what Maurin called "a philosophy so old it looks new." They are committed to putting forth a positive vision (seeking, as Maurin said, "to announce not denounce"), expressing and embodying a way of life that seeks to elude, contest, and withstand the trappings of modernity, in the attempt to create "a world in which it is easier for people to be good." They would welcome your partnership, encouragement, critique, good humor, and help.[7]

6   Jim Forest, "The Catholic Worker Movement," http://www.catholicworker.org/historytext.cfm?Number=78
7   For more about the Catholic Worker movement, including a listing of communities throughout the world, see: www.catholicworker.org

PETER MAURIN

Marice Sariola

Regard For The Soil
Peter Maurin,
From *Easy Essays*

1

Andrew Nelson Lytle says:
The escape from industrialism
is not in socialism
or in sovietism.

2

The answer lies
in a return to a society
where agriculture is practiced
by most of the people.

3

It is in fact impossible
for any culture
to be sound and healthy
without a proper regard
for the soil,
no matter
how many urban dwellers
think that their food
comes from groceries
and delicatessens
or their milk from tin cans.

4

This ignorance
does not release them
from a final dependence
upon the farm.

A selection from the Aims & Means of the Catholic Worker (reprinted in The Catholic Worker, May 2012)

…we advocate:

—A decentralized society, in contrast to the present bigness of government, industry, education, health care and agriculture. We encourage efforts such as family farms, rural and urban land trusts, worker ownership and management of small factories, homesteading projects, food, housing and other cooperatives--any effort in which money can once more become merely a medium of exchange, and human beings are no longer commodities.

—A "green revolution," so that it is possible to rediscover the proper meaning of our labor and our true bonds with the land; a distributist communitarianism, self-sufficient through farming, crafting and appropriate technology; a radically new society where people will rely on the fruits of their own toil and labor; associations of mutuality, and a sense of fairness to resolve conflicts.

For the full list of Aims and Means of the Catholic Worker, see:
http://www.catholicworker.org/aimsandmeanstext.cfm?Number=5

EDITOR'S NOTE Andre Lytle was part of the "Southern Agrarians" literary movement which advocated for an increasingly vibrant agricultural economy and culture. Check out his book *I'll Take My Stand*

EDITOR'S NOTE The National Catholic Worker Farm Gathering is happening from the 20th–23rd of February 2015. For more information see: http://www.thegreenhorns.net/ailec_event/national-catholic-worker-farm-gathering/

Dorothy Day

November

Indigenous

Chapter 11

Potatoes and Patterns:
A Tapestry of Beauty and Diversity

Sonja Swift

In the highlands of Peru not far from the ancient Incan city of Cusco there is a very special kind of park. Named simply after a familiar food native to the Andes, it is called Parque de la Papa, or the Potato Park. Nature parks or wilderness corridors are usually viewed as places fenced off and separated from human use, but this park doesn't bar people. Rather it celebrates the role people play as stewards in caretaking landscapes.

Parque de la Papa is a 9,000 hectare Indigenous biocultural territory comprised of five Quechua communities: Sacaca, Chawaytire, Pampallaqta, Paru Paru and Amaru. The park has been officially registered and granted park status by the Peruvian government, thereby protecting the land from mining, logging and other outside business interests. Parque de la Papa is located some 40 kilometers northeast of Cusco and neighbors the Sacred Valley of the Incas, the gateway to Machu Picchu.

Park status was designed to protect the land, but also to make an important point: that agro-ecological parks and food sovereignly corridors deserve equivalent recognition and safeguarding. As described on the Parque de la Papa website, "A considerable part of the world's biological diversity is found in territories whose ownership, control and use correspond to indigenous and local communities, including nomadic peoples. Nevertheless, the fact that these peoples and communities conserve many of these sites, actively or passively, through traditional and modern ways, has been largely ignored in official conservation policies."

While the park was officially formed in 1998 with the help of Cusco based NGO Asociación ANDES[1], it wasn't until 2004 when ANDES succeeded in returning not only native potato varieties from the International Potato Center (a root and tuber research-for-development institution) but also the rights to those potatoes. Having reached an agreement with the IPC, ANDES was able to repatriate these potato strains. This homecoming re-forged a connection that had been lost. In the words of ANDES Director Alejandro Argumedo: "In looking back that is what makes me happy, that the humble potato can make that difference."[2]

Efforts to repatriate the potato continue and today the communities are growing a total of 1,342 different varieties of potato plus some 83 varieties of mashua and other Andean roots. Estimates about the total number of existent potato varieties place around 4,000 of them in the Andean highlands[3]. Thirty varieties of rainbow potatoes had recently been repatriated when I visited in March of 2014, and in one small greenhouse over 400 new varieties were being tended for planting in the fall.

---

1   See: www.andes.org.pe
2   Personal interview with Alejandro Argumedo 03/25/14
3   http://www.cipotato.org/potato/native-varieties

The communities manage several collective projects including a medicinal plants group, handicraft collective of traditional weavers, the potato guardians, and the gastronomy collective who also run a restaurant. There is a common fund, toward which each organized group donates 10%—binding their work with one another. This is like the rotational planting system, which also helps to unite the community.

Parque de la Papa is the work of the communities. They decide how things should be done while ANDES helps to assure their rights are protected with clear agreements. In a recent conflict, the National Institute of Agricultural Research tried to lay claim to fifty potato varieties. Together ANDES and Parque de la Papa protested and won, preventing more seed-patenting thievery.[4] In 2011 the Peruvian government signed a 10-year moratorium against GMO's and Cusco has been successful in asserting itself as a GMO-free region, but threats persist.[5] As one of the Potato Guardians stated, "Potatoes are not just seeds but are also linked to our struggles, against transgenics (GMO's), and in reclaiming potato varieties."[6]

Some approximations say that in the next 40 years the potatoes will reach the mountaintops because growing zones are moving so rapidly. The acceleration of change is being witnessed up close here, and people are taking detailed account of the changes in their everyday interactions with the land. Diversity is their resilience strategy and they are actively working on breeding new potato varieties for lower elevations as well.

There is a sense shared by many allied with Parque de la Papa that our current ecological crisis is happening because of a lack of spiritual and land-based values that embody respect for the living planet. We evolve (or devolve) with the food we eat and the water we drink. We are molded by the landscapes we inhabit,

to destroy the ecosystems that sustain us is to destroy our very potential for complex intelligence. This is reflected in the integrity, care and thoughtfulness of the communities in Parque de la Papa. Their landscapes, like their handmade weavings and vibrantly colored ponchos, are tapestries of beauty and diversity reflecting a worldview of interconnectedness.

Weaving itself is a paradigm. It is mirrored in the landscape, which looks woven, as opposed to the feudal version of big square blocks sliced apart and divided like prison cells. This paradigm has a lot to do with beauty, and with living life guided by beauty. There is nothing more beautiful than a tapestry of diversity, a mosaic of fields growing over a thousand varieties of potato, alongside corn, amaranth, and quinoa blooming iridescent. Like intact prairie, old growth forests, or coral reefs, diversity is where all textures and colors come together, where life pulses and shines.

In beauty I arrived at Parque de la Papa, with a handful of flower petals tossed over my head ceremoniously. Upon departing I was given a hand woven maroon and jade green scarf, placed cordially around my neck by Ricardina Pacco Ccapac the first woman to serve as president in her community. In beauty I was welcomed and in beauty I was bid farewell, continuing the tapestry that is Parque de la Papa.

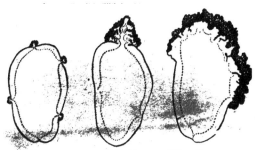

Potatoes Cut in Half Showing Wart Disease.

4   http://www.biocultural.iied.org/sites/default/files/INIA%20final.pdf
5   http://www.lab.org.uk/peru-a-10-year-ban-on-gmos
6   Personal interview, Potato Park 03/25/14

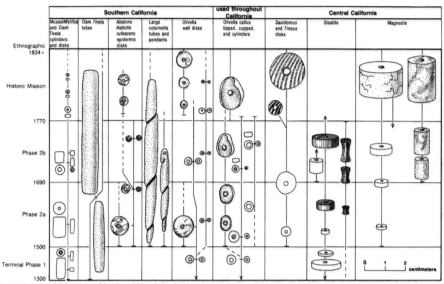

Fig. 2. Common California protohistoric and historic beads.

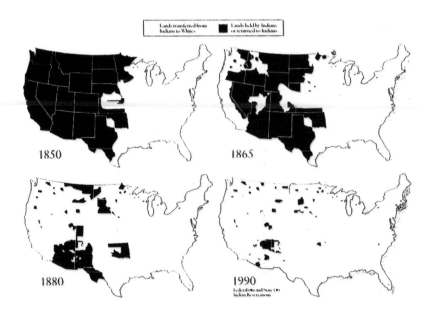

Cultivate With Prayer:
Sacred Foods Of The Native
American Church

Katherine Millonzi

Founded in Oklahoma in 1918, the Native American Church (NAC) is the most widespread religious organization among Native peoples of Northern America. The tipi is one of the NAC's places of prayer; tipi "meetings" are sponsored events, held in honor of a particular community member calling for support. The tipi is a considered to be an altar for communion with the sacred. These overnight rituals conclude with an offering of four sacred foods, which are brought into the tipi in the morning, and shared by all participants. The cultivation, preparation, and consumption of these sacred foods—water, corn, meat, and fruit—are all undertaken with intention and prayer. There are specific meanings for each of the sacred foods, the water, the fire, and all the sacred instruments used within the ceremony. What connects these teachings is an approach to food that emphasizes balance and health, and honors the elements that connect spirit and body.

Each of the four sacred foods is a prayer, symbolizing the four sacred directions—North, East, South, and West. The foods are passed around the tipi for each to partake. By passing around the circle, all four directions are honored. The circular tipi mirrors the quadrants of a human life and the foods, according to tribe, correspond to the four aspects of a being:

Mind, Body, Spirit, Soul. These foods are invocations of the natural forces that come to aid in the gathering's intention.

At midnight a man carries in the Water, which has been collected in containers ritually consecrated to the task. The Water, itself considered a feminine element of memory and purification, circles the fireplace in the center of the tipi. Through this meeting of Water and Fire, some members of the Church believe a prayer is made for the balance of the divine feminine and masculine energies in the great circle of life.

The bowl of corn that follows is hand-ground, and often cooked with butter and maple syrup into a sweet mush. Older strains of maize are favored, and seed varieties are cultivated and passed through home gardens, retaining flavor complexity and genetic adaptations to specific climates. It is understood that landrace species become attuned to the people that cultivate them, and vice versa. The seeds, planted with a deer heart and tobacco rolled into last years' shucks, are tended lovingly into stalks and harvested as colorful ears. Cultivated with a prayer, for use in prayer, corn is recognized and respected as a feminine food, connected with our minds.

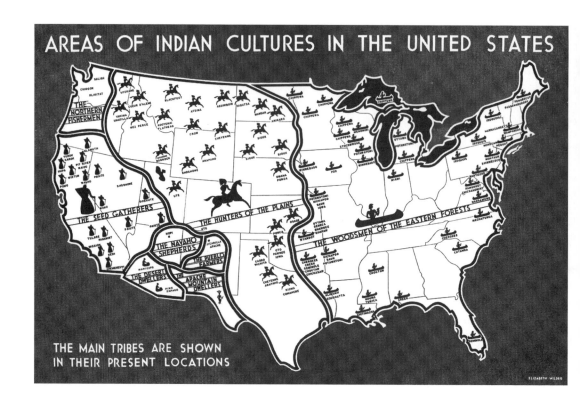

AREAS OF INDIAN CULTURES IN THE UNITED STATES

THE MAIN TRIBES ARE SHOWN
IN THEIR PRESENT LOCATIONS

The animals, it is believed, offer themselves for use in ceremony. Typically men carry out the hunt, but some tribes have female hunters. Animals hit by a car, or otherwise injured violently, are taboo to consume. When an animal is killed, corn and tobacco are placed in its mouth as a sign of thanksgiving and reverence. After the primary breakdown of the carcass, no knife is used in further preparation of the sacred meat. Meat is browned, slow braised, and then shredded into a bowl by hand. Occasionally pine nuts, walnuts or pecans and cranberries are mixed in. In some regions of Canada and the Pacific Northwest, salmon is used as the sacred meat. No part of the animal is wasted—many ritual uses exist for each sacred organ and part. While everyone in the ceremony eats together, some people who have taken spirit relations with certain animals do not eat their spirit kinfolk. Associated with the masculine, the sacred meat is connected to our bodies.

The bowl full of berries, a welcome sweet after a long night of prayer, are wild and indigenous—often choke cherries, cranberries or blueberries. The act of picking is undertaken with prayer and the fruit pickers bless themselves and the fruit four times each with cedar smoke. Depending on the tradition, the sacred fruit is connected to either the child or the elder, and associated with our spirits.

The tipi ceremony offers a direct encounter with the spirit realm, and the ceremonial sacred food is considered profound medicine, capable of curing any ailment. As each food brings its own spiritual sustenance and healing, the food itself can be directly asked for help. Each bite—usually four of every dish—reaffirms embodiment and life, and grounds the ceremonial space. As the body is nourished, the will and honor to live a new day are made conscious again.

Joseph Campbell has said the only "in-group" is the planet. There is nothing unaffected by the great scales of planetary balance. When food is approached without due gratitude and attention, when reverence is breached, it reflects and perpetuates the wider imbalances that prevail in our relationships to the natural elements. Access to traditional sacred foods is threatened in many cultures. Mass slaughter of the American Bison, for example, resulted in the near extinction of the herds in the 1800's. Today, the purity and safety of our waters are endangered by hydro-fracking and pollution. The emerging long-term impacts of transgenic (GM) sweet corn on the environment, health, and traditional landraces are dismal. Greed and arrogance underpin this disrespect for our sacred common resources.

The Native elders teach that gratitude is sustenance. The practice of gratitude feeds our cells with softness and nourishes us with humility. The Earth creates herself in forms that we can ingest at her splendid table. By cultivating homage, in ceremony and at every daily meal, we are given the opportunity to return her hospitality.[1]

---

1 Thanks to Water Woman Ann Rosencrantz, member of the Native American Church and Director of the Council of 13 Indigenous Grandmothers, for passing on her teachings, learnings and love.

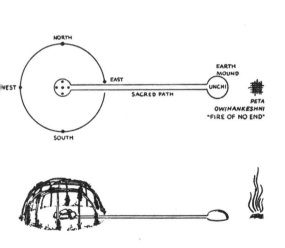

The *Inipi:* "Purification Lodge"

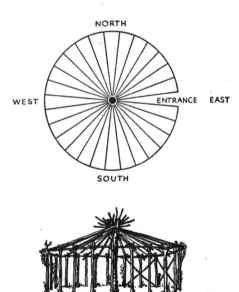

The *Sun Dance Lodge*

EDITOR'S NOTE Lodges used by the Oglala Sioux for two of their sacred rites—The Inipi, or Rite of Purification, and the Wiwanyag Wachipi, or Sun Dance.

Indians Buffalo-hunting in Masquerade.

Indians Elk-hunting in Masquerade.

EDITOR'S NOTE These are European depictions, inauthentic, seeking to explain native hunting techniques. Europeans and contemporary Americans remain confounded by indigenous prowess and virtuosity with simple, elegant tools and weapons.

Certified Campesino:
Learning From Agroeconomical
Advances In Peru

Kat Shiffler

Every September, around half a million foodies gather in Lima, Peru for Mistura—the country's showcase of gastronomy and haute cuisine. Like in the United States, Peru is in the process of a heavy re-emphasis on food traditions. In the last decade a number of alliances between chefs and small-scale producers have emerged. At Mistura, farmers, fisherfolk and pastoralists are center stage, representing an unfathomable biodiversity from the Amazon to the Andes to the Pacific Coast. Visitors to the Gran Mercado can see, sniff and sample everything from a rainbow of native potatoes to a scarce honey from stingless, jungle bees.

In the middle of the crowded marketplace is la Asociacion Nacional de Productores Ecologicos del Peru/ the National Association of Ecological Producers of Peru (ANPE-PERU), a small-farmer group comprised of over 12,000 members from 22 different regions of the country. Their stand is stocked with a multitude of quinoa, tubers, colorful corn varieties, spices, jams, fruits, garlic, and heritage seeds. It's all marketed with a snappy, green label that reads Frutos de la Tierra, Fruits of the Earth—a collective brand of ecological goods characterized by campesino, or small-holder agriculture. In Peru, it is estimated that for every ten food items eaten daily, seven are produced by campesino agriculture.

Frutos de la Tierra, although a new venture, represents long-standing trends within Latin American agroecological movements—what academics like Miguel Alteri and Victor Toledo call a "new agrarian revolution." The movement promotes local/national food production through access to land, seeds, water, credit and local markets and the creation of supportive economic policies and market opportunities for agroecological innovation. At the base of this revolution are small-scale farmers organized in horizontal networks and social processes that emphasize self-reliance and value community involvement.

Currently, Frutos de la Tierra includes more than 200 products from all over Peru. To participate, a farmer's products must meet three main requirements: 1) it must be produced in an ecological manner on a "family-scale"; 2) it must represent the in-situ conservation of biodiversity via smallholder agriculture; 3) the producers behind the product must work cooperatively. In addition to being absent of agro-chemicals and GMOs (Peru was the first country in the Americas to ban genetically modified foods), the brand symbolizes an underlying philosophy of long-term food sovereignty via economic sustainability, local markets, cooperativism, and the protection of biological resources by the farmers themselves.

To assure that this is the case, ANPE-PERU created a Participatory Guarantee System (PGS)—a system of quality assurance that's an increasingly popular alternative to third-party certification. Aimed at bolstering local economies through direct, transparent relationships, producers and consumers establish their own context-specific guidelines for agroecological production systems. The International Federation of Organic Agriculture Movements (IFOAM), among others, is a strong proponent of these systems around the world.

Cooperative enforcement is based on verification among producers, academics, and consumers. In the case of Frutos de la Tierra, world-famous chefs partner in a cocinero-campesino alliance. The certification process starts with a foundation of trust. And in the spirit of food sovereignty, this participatory process assures that the power to certify is not concentrated in the hands of a few, profit-centered bodies, but in a cooperative process. This concept is central to Frutos de la Tierra's added-value; an emphasis on not only regenerative ecological processes, but social ones as well.

But can ecological peasant agriculture "feed the world?" While the masses were chowing down at Mistura, hundreds of researchers, professionals, students and farmers on the other side of Lima pontificated on exactly that question at the 4th Latin American Congress of Agroecology.

Attendees from 30 countries shared their research and their experiences. Cubans showcased their intensive and infamous urban systems; Colombians reflected on the ongoing farmers' strike that paralyzed the country; others talked about biological control in banana plantations in a specific corner of Venezuela.

A common theme was Latin American Agroecology's deep ties to social movements—specifically the transnational agrarian movement La Via Campesina, of which ANPE Peru is a member. On the explicit political agenda: dismantling the industrial agrifood complex, restoring local food systems, and implementing genuine agrarian reform via access to land, water, and seeds. And this could be by grassroots revolution if not by enlightened public policy.

According to the Latin American experts at the Congress, campesino agriculture accounts for more than 50% of global food production, and just over 20% of arable land use—meaning industrial agriculture, although it occupies 80% of the world's best soils, produces only a third of our world's food. These agroecologists operate from the premise that small farmers are the solution to food security. They're are also central to climate change adaptation, rural development, and preserving biodiversity. Frutos de la Tierra is a good example of this academic vision in action: going beyond certified organic, beyond farmers as quaint symbols of gastronomy and culture, to create a participatory producer-consumer movement that values and promotes the role of small-scale farmers as essential for future food sovereignty.

Josef Beery

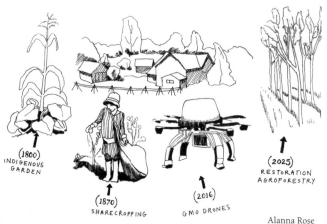

(1800)
INDIGENOUS GARDEN

(1870)
SHARECROPPING

(2016)
GMO DRONES

(2025)
RESTORATION AGROFORESTRY

Alanna Rose

Selected Principles of the Agrarian Trust

OPPORTUNITY

Agriculture has been a pathway of opportunity and upward mobility for many generations of Americans. Anyone highly motivated, including new immigrants, socially disadvantaged, minority and female farmers should be able to climb the ladder from farm worker to farm manager to farm owner. Efforts to expand access to land, capital, technical assistance, etc. must include those who currently labor in the fields, and create pathways to ownership and co-ownership. Farm service providers are crucial allies in supporting disadvantaged growers, and protecting their interests.

AFFORDABILITY

Farmland should stay affordable from generation to generation, whether inside the family or outside of it. Often, slowing down the transaction makes transition to a farmer-owner possible. This can be accomplished in many ways, including lease-to-own, contract-sales, state-subsidized land banking mechanisms, affordable lease agreements, conservation easements with affordability clauses. We must increase access to credit, alternatives to credit and more progressive credit for farm purchase and capitalization.

PROTECTED IN PERPETUITY

Farmland should be protected for production in perpetuity, and farmed using sustainable methods that preserve the public trust. Farmers must be able to invest in the health of the land, protect vulnerable areas, and improve wildlife areas, habitat, water quality. Land that is protected must stay in farming, farming for food, and may not be converted to equine or purely recreational use.

AGRARIANTRUST.ORG/PRINCIPLES

LAW OF CONTRACTS

A CONTRACT IS AN AGREEMENT OF TWO OR MORE PARTIES, by which reciprocal rights and obligations are created. One party acquires a right, enforceable at law, to some act or forbearance from the other, who is under a corresponding obligation to thus act or forbear.

Generally speaking, all contracts which are made between two competent parties, for a proper consideration without fraud and for a lawful purpose, are enforceable at law.

TO THE CREATION OF A VALID CONTRACT there must be:

1. Precise agreement. The offer of one party must be met by an acceptance by the other, according to the terms offered.

2. There must be a consideration. Something of value must either be received by one party or given up by the other.

3. The parties must have capacity to contract. The contracts of insane persons are not binding upon them. Married women are now generally permitted to contract as though single, and bind their separate property. The contracts of an infant are generally not binding upon him, unless ratified after attaining his majority. The contracts of an infant for "necessaries" may be enforced against him to the extent of the reasonable value of the goods furnished. It is incumbent upon one seeking thus to hold an infant to show that the goods furnished were in fact necessary to the infant, and that he was not already supplied by his parents or guardians.

4. The party's consent must not be the result of fraud or imposition, or it may be voided by the party imposed upon.

5. The purpose of the parties must be lawful. Agreements to defraud others, to violate statutes, or whose aim is against public policy, such as to create monopolies, or for the corrupt procurement of legislative or official action, are void, and cannot be enforced by any party thereto.

CONTRACTS IN GENERAL ARE EQUALLY VALID, WHETHER MADE ORALLY OR IN WRITING, with the exception of certain classes of contracts, which in most of the States are required to be attested by a note or memorandum in writing, signed by the party, or his agent; sought to be held liable. Some of the provisions, which are adopted from the old ENGLISH STATUTE OF FRAUDS, vary in some States, but the following contracts very generally are required to be thus attested by some writing:

Contracts by their terms not to be performed within a year from the making thereof.

A promise to answer for the debt, default, or miscarriage of another person.

Contracts made in consideration of marriage, except mutual promises to marry.

Promise of an executor or administrator, to pay debts of deceased out of his own property.

Contracts for the creation of any interest or estate in land, with the exception of leases for a short term, generally one year.

Artist-Educator Farmer:
Walking With Aimee Good

Ernesto Pujol

Aimee Good flows between her roles as Director of Education and Community Programs for The Drawing Center in New York, and managing an organically certified garlic farm in Maine. She describes it as passing through a membrane, from a public to a quiet place, and she does it well. During a recent foggy morning, we took a long meditative walk through Prospect Park after she and her husband delivered their nine-year-old daughter to a nearby public school.

ERNESTO PUJOL What was your training?

AIMEE GOOD I received a BFA in sculpture from the Massachusetts College of Art, and an MFA from the Milton Avery Graduate School of the Arts at Bard College. But I come from a long line of potato farmers in rural Northern Maine. Growing up, I spent a lot of time outside, in the fields, in the woods. Much was asked of me at a young age; I was given a lot of agency in the family farm. My hands were always employed with farm tasks: picking, sorting, pulling; and with interior domestic tasks: cooking and baking. My hands went through a lot, learned and know a lot. My father let me into the larger picture, riding the tractor with him. He pointed out the arc of the horizon as it met the rolling edge of fields and trees. I saw color, the cut of furrowed lines drawn upon his freshly turned soil. He taught me to see.

EP You experienced making with passion long before art making.

AG Farming is hard work; you have to find your grace in it. I learned to meditate when I was seven years old. I learned what it was to work in the sun alone, engaged in repetitive labor, in wholesome patterns. I grew up in a family where, whatever you took on, you applied yourself with your full being. Farming is not a job; farming is a life. And there is a lot of emphasis on how things look, a strong relationship to beauty. My mother, who has a lightness that I admire, has always said that I was very serious early on. I often did the work that no one wanted to. Much later, artwork would become the natural extension of what I was already doing as a child, holding people's hearts.

EP How did you transition to your current artwork?

AG One night I experienced a vivid dream in which I found myself walking with animal guides across a landscape and came across living bones. They looked at me but it was not my time yet. I was meant to walk on to witness the passage of time. So I began to make installations of rib cages from bulls, horses and goats. I poured and spilled milk that stained and molded, creating images. I made a heavy dough dress and wore it. The stories came from the farm. My hope was to bring people into an installation space that I created and spark something—trigger memories. I worked with discarded materials from a butcher, such as veal tongues. I cast body parts; I made videos; I experimented a lot. I made work as ugly as possible. My last name is Good, and sometimes I got as far from good as I could, to look at it, so I could see it; to come back and know what it meant to be good in the world.

EP It feels like work from memory—healing memories?

AG I think of my pioneer ancestors who came to a place, pierced the earth, and worked very hard to make it. But with that came toughness. There was a positive: making a dream happen. There was darkness, a constriction: the pressure to conform and not throw the system off by being different. There was shared labor, but women's opinions did not count for much outside the home. So I inherited a thick and heavy mantle from the women in my family. I carried a deep sadness that I had to work through.

EP How has it been to live and work in New York City?

AG I am attracted to opposite poles: from a completely rural landscape to a completely urban one. I moved to the city after a fellowship at the Irish Museum of Modern Art in Dublin, which marked a closure. I made a sculpture of bread, passed it around for the audience to hold and eat, if they wished, and silently walked away. In NYC, I initially experienced a shrinking of my gestures. I worked for a while for Martha Stewart, who even filmed a program in our family farm. Giving birth to my daughter brought me to a crossroads. Priorities were reset. My creativity, out of necessity, shifted. I began to contemplate inviting the spirit of collaboration into my life. I missed the art community and began to work with the education departments of the Museum of Modern Art and the Cooper-Hewitt, National Design Museum. I had visited The Drawing Center throughout my undergraduate and graduate school and, when a six-month position opened there, I applied. That was six years ago.

Aimee Good

EP So, it has been a learning experience, yet again.

AG New York has taught me how to find the center of my power as a human being. I am eternally grateful for that.

EP Do I also hear a new artist-residency in the making?

AG There is great powerful energy in northern Maine. I have always thought of creating an artist residency program located within the context of an active farming community. I register a level of authenticity in communities where economies are established in relation to natural resources. The best art residencies I have ever experienced where those in which I was asked to step outside of my comfort zone and present my work to a community—to give back to a community. I plan to build a mobile pizza oven and start hosting community dinners in my hometown, to build awareness and solidarity for the development of more local and regional food systems. I want these art residencies to be about understanding the interconnectedness of everything. My organic garlic is not about me: it is about all the people who helped produce it and will be fed by it.

EP Art is like good bread.

AG Yes! I seek conscious choices, a knowing and seeing even if those choices are not always understood, accepting your endings and daring beginnings. I gave myself the gift of art making. I continue to do long, collaborative, large-scale projects that merge public participation, in terms of art as social practice, to allow multiple points of access. My Maine survival skills and understanding of the facets of local communities have served me well living here. But I also choose to return to farming, to collaborate with my father and mother to reclaim the future. Family farms are dying. I have a unique perspective to put in front of audiences. Reclamation is big with me. I like the Native American notion of one running ahead, to make sure that the way is clear for the rest of the tribe.

Aimee Good

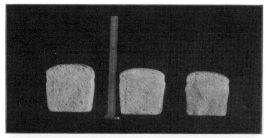

Aimee Good

To make bone broth

Put bones in a big big pot
the biggest you have
put it on the stove
boil it, then cook gently for a while
until the flesh is all off
optionally, remove some fat for
future or other use.

Other ingredients to consider:
bay leaf, oregano, thyme, other
herbs, salt, pepper to taste.

Bone broth is very nourishing.

Pork bones for ramen,
Chicken bones for sick bunnies,
Beef and lamb broths for libido.
Allegedly.

For more information on
this topic check out:

www.threestonehearth.com

The Radical Homemakers,
Shannon Hayes

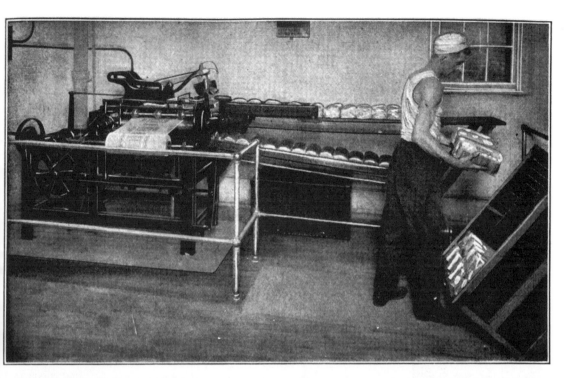

Island Ethics

Catie Hannigan

you are an island
that dreams fathomable dreams
as perfect as color
as gold tobacco eyes
that doesn't care about
a gentle beach reed is your harm
the travelers   the transient
the historians who spit on the ground
everyone is looking at your aloneness
you will rise and beat
a wave that is everywhere
and unbounded
the sand spirals above
your body's sweetness
time is an open window
where you
escaped

Blue Wildrye

you are on an island
the sky silky as french
as golden tobacco balm
the beach reeds have grown
ten feet high and harsh enough
to scratch the skin of everyone
trying to move through
and reach the sea
scrapes and bruises mean
the other side is the sea
waves are sacred and profane
they beat
and beat
and beat
they spit out jellyfish still glowing
upon land that is everywhere
and no one's

Mountain Brome

December

REAP THE BOUNTY THAT YOU SOW

Shifting

Chapter 12

Finding The Magic and Manure
in One Family's Farming Chronicles

Lori Rotenberk

It's a Sunday ritual for Henry Brockman to drive down to the fields and walk the rows, taking notes about the growing season in one of four notebooks he keeps in his truck. Work boots pressing into the soil, braced for Central Illinois' moody weather, he walks because the paces help him think, sliding his pen over to the book's margin to scrawl a thought or a line of poetry.

They're "little encapsulations of a time and place," he says. "A farmer's life is about knowing what time it is."

Dozens of such notebooks fill boxes at his home, which anchors Henry's Farm, a 12-acre Eden located in the fertile bottomland of the Mackinaw River Valley outside of Congerville, Ill. And Henry is not the only one of

Alexandra Sing

six Brockman siblings who has picked up the pen. Most of the family not only farms, but also writes—a trait, Henry's sister Terra says, that is "genetic."

The Brockmans matter in the shifting world of farming because they are taking the time to catalog their work, their days, their thoughts. Well-educated and multi-cultural, they write about the challenges of the seasons, the hard realities of farming organically, and the turmoil of climate change. If you're pondering the ag life, or if you just want to understand in a deep way where your food comes from, their works are a must-read. Ecologist Sandra Steingraber likens Terra's writing to that of Wendell Berry.

Here, have a taste, from Terra's James Beard Award-nominated book, *The Seasons on Henry's Farm:*

Last week I sliced my finger while cutting the turnip greens during the harvesting marathon, but didn't realize it until I returned home hours later because my fingers had been numb and almost bloodless. If you put yourself into the muddy boots and frozen feet of a person who has been doing hard manual labor out in the fields every minute of daylight for the past eight months—through the brutal heat of summer, through downpours, through freezing rain and wind, hour after hour, day after day, month after month—then you can imagine how pleasant winter begins to look.

Alexandra Sing

Jotting down their thoughts was originally a way to bring CSA and farmers market customers the backstory belonging to the food on their plates. Terra, who has a degree in English literature, initially wrote a two-page newsletter she calls "Food and Farm Notes" that she would hang from the rafters of their stand. Filled with facts, literary references, and tidbits about the produce available at market that day, it garnered a cult following. Copies would disappear within an hour. She began sending it by email, and that eventually led to her book.

The next generation has inherited the writing gene as well. Henry's 21-year-old daughter, Aozora (Zoe), wrote poetry and essays about growing up on the farm in a chapbook, *The Happiness of Dirt*, and is now studying creative writing at Northwestern University.

But like many young people raised on farms, Zoe is unsure whether she'll dedicate her own life to agriculture. "I love having grown up on a farm, and didn't know how different and wonderful it is until I went away," she says, but adds: "It's becoming harder and harder to be a farmer."

"Climate change is very hard on us—each year is different and very chaotic," Zoe continues. "This year we had rain, but the past years it was super dry. Dad had to irrigate and we'd wake up every three hours and place drip tape on different rows."

Here's a bit from her essay "Kill or be Killed":

No rain for weeks made cracks appear that sliced the soil into great slabs, heavy as rock, and those I moved— teeth grinding slow to keep from thinking of the rays of sun that lit my back ablaze and how my fingertips felt ripped open each time I dug at the coarse soil, in search of smoothness.

While Zoe contemplates what to do with her life, she and her father have decided to write another book, cataloging the two different generations examining their farming life.

Next for Terra, 55, who founded and heads the Land Connection (www.thelandconnection.org), a nonprofit linking available farmland with new farmers, is a "very approachable" book on soil. "Soil is precious and complex, like a giant internet," she says. "Civilizations have risen from the soil and they've fallen due to agriculture."

Come January, Henry is taking a year off from farming to live in rural Japan with Hiroko. There, he says, he will begin sifting through the notes taken in the fields over the past 22 years. The rest of the family will keep Henry's Farm in operation while he's away.

Upon hearing the news, one regular farmers market customer announced on Facebook: "My farmer is taking a Sabbatical!" "If every 21st century American had a relationship with a farmer, much as we have a preferred auto mechanic, we would all be heedful of what we eat and the labor that it requires to produce it," she wrote. "Maybe we would eat less, enjoy it more and be the better for it."

You can find the Brockmans online at blog.brockmanfamilyfarming.com/[1]

GODSPEED THE PLOW

Though the wealthy and great
Live in splendor and state
I envy them not, I declare it
For I grow my own hams
My own ewes, my own lambs
And I shear my own fleece and I wear it

For here I am king
I can dance, drink and sing
Let no one approach as a stranger
I'll hunt when it's quiet
Come on, let us try it
Dull thinking drives anyone crazy

I have lawns, I have bowers
I have fruits, I have flowers
And the lark is my morning alarmer
So all farmers now
Here's Godspeed the plow
Long life and success to the farmer

---

1   This piece originally ran on Grist, www.grist.org

Black Walnut

Bobby Losh-Jones

Most of the trees produced our first year.
How exciting to spot the
rotting green and black golf balls
the size of your fist
along the thicket in the hedgerow.
We kicked off their musty black rot
to reveal wrinkled shells inside.
They dried a day or two on the front stoop.
We cracked them with a borrowed hammer—
a lot of work
for a few shards of nut.
They tasted overwhelming,
like eating a flower
you are not allowed to smell
because its fragrance
is too intoxicating.
I could eat only one.

We vowed to collect more—
gifts from a tired land—
and store them through the winter.
We never made it to that part of the list.

Next fall, worn down from a year of work,
we walked the hedgerow searching.
There were no nuts to be found.
We came to know the walnut trees
by their bare branches
succumbing to thicket.

There are plans to cut back the thicket,
strip the bare branches,
and take the rest to a workshop miles away
where they will become
a table, cabinets, a chair,
an expensive chest.

Today we learned the secret
of the third person living in our house
growing like a seed inside you,
still smaller than a walnut.
I am 27 years old, you are 26.
Black walnut trees can take over
thirty years to produce.
I ate only one.

Next fall, in the empty hedgerow
we will dig holes
and plant young seedlings
and watch them grow.

Black Walnut

Letter To Myself

Byron Palmer

When I was 26 years old in 2007, I decided I wanted to quit my job making documentaries about agriculture and become a farmer. Now, seven years later after a very circuitous path to becoming a grassland manager, I sometimes wish I could go back and give myself a little advice.

Dearest Byron Self,

This is a letter from you, in the future. Please pay attention, I have some advice. I know you are excited about farming, and the nurturing, collaborating, exercising, thinking, and tending the earth that is part of it.

You have romantic notions about growing food as a career, and it will be a lot different than you think. To make any idea happen you must get moving. You have taken gardening classes, read books, home-scale gardened for a few years, gone to farmers markets, but you need to get your hands dirty as soon as possible.

Keep your current job and moonlight as a volunteer on a local farm on your days off. Find a farm where you can do a lot of hard work, and if the shift is short ask to stay for the full 9 hour day. This is not considered overtime in agriculture. Do this for at least 6 months, in

all extremes of weather. If you're still excited, it's time to start looking for an apprenticeship. There are plenty of apprenticeship programs out there.

Develop clear goals about what you are looking to learn. At least one of your goals should be to see how the lifestyle of 50 or 60+ hour weeks suit you. Keep your focus on learning skills while you're there, but I recommend sinking into the pattern of long hard work and being honest with yourself about how it feels to you as a way of life. It is not as important to be skilled in this endeavour as it is to know this is the right endeavor to be skilled in.

If you find that you do indeed love the pace and work of farming or ranching, it's time to consider what kind of operation you want to have—one you own or one you work for. If you want to work for someone else you must develop your technical skills and farm craft and be square with making around $12-$15 an hour for the rest of your life. If you want to run your own operation, you will need to develop your business skills as much, or more, than your farming skills.

If you like plants and farming, you probably are not a big fan of business. Well guess what, if you want to run your own farm or ranch you will be a business person. It's easy for new farm start-ups to go out of business, and it often has nothing to do with the quality of their product. Most ranches lose money every year. If you decide to start your own, you will be running a business in one of the toughest industries there is.

But there are farmers and ranchers out there making money, having fun and healing the earth. If you want to go into business for yourself, I recommend starting by finding some of these thriving people and working with them. Learn everything you can about their business strategies, but also as much as possible about their personal strategies for staying happy and healthy.

Eventually you might be ready for your own operation, and for this you'll need land. And in California (which is where you'll stay, friend) you can forget about buying land. I know this has always been your dream, but land ownership is complicated. Land is layers of value— views from it, wind over it, development and mineral rights to it, infrastructure across it, grass on it. When you buy land you are paying for all these layers, but only extracting value off of a few of them. You're no longer in the farming or ranching business, you're in the lvand business. If you want to avoid this you will want to lease so that you can pay only for the layers you need for your operation.

If you don't want to lease or buy you can consider other ways to find land for your use. Here in California there are tons of opportunities for creative tenant agreements—from ranchers wanting non resource competing enterprises to compliment their pasture (read chickens) to new land-owners with an unused acre or 3 to spare. By doing your research and picking carefully it is very possible to find ideal land to start your operation with little to no cost.

Once the land thing is figured out you'll have to attend to the particulars of the day-to-day operation of your enterprise, its distribution methods and scale. I can't lay out all the answers here, but it will be important to do enough research to thoroughly understand the costs and regional influences (weather, distance to markets, ecology, topography) on running your operation. You'll also need to decide if you want to operate on a direct-to-consumer or wholesale distribution model. Your distribution method will dramatically affect your overhead cost structure. Pay attention to detail.

Lastly it is important to begin with the end in mind. Do you want to buidl a business, build a herd, build equity and infrastructure? Or do you want to manage someone else's land? You might decide it's better to run someone else's operation, garden at home, or sell veggies to friends and family. That's ok, and it's important to understand that there are dozens of legitimate paths to take, dozens of ways to contribute to the food system and nurture the land. Remember that the opportunities are endless and the world needs you. Just listen, and follow your intuition.

Your Friend,
Your future self

Katelyn Hale *Pasture Cycle*

Selected principles from On The Commons, a gathering to discuss the state of the land and lay out some guidelines about what is needed to precipitate change in issues of use and ownership of farmland:

When we buy and sell land we are really buying and selling certain rights of use to the land, rather than the land itself. And these rights are always balanced by responsibilities. Therefore, having the right to a certain piece of land should always come with an obligation to practice social, economic, and environmental stewardship.

Land should not be treated as a commodity or speculative asset.

We need to restore a context to land use that is relational, long term, and responsible.

Farmland should be liberated from industrial production and large scale consolidation.

Land should be reconnected to livelihood and community and provide a sense of place.

We can work from an understanding of "right relationship" among animals, nature, and people with the understanding that nature is the ultimate steward, a source of guidance and knowledge.

Mission investing is a profound paradigm shift. We need to dispel the mythology of wealth—what we create as a community—as only money and capital.

Leaving the design of a new food and land system to existing bureaucracies and power structures will not lead to the new vision.

We need a range of strategies and new models to create change, including legal structures, contracts, policy, public support, investor models, etc.

We need to pay attention to both land retention and land availability for new farmers.

There is a need to create a culture that allows us to foster and operate out of a different land paradigm.

We need to activate what is not being used, what is laying fallow as unseen potential. We are in abundance, under-using, and not reusing.

We need to cultivate a different investment mentality and practice with regard to land. Slow money is not only about return, but also about maximum community benefit.

PRINCIPLES FROM WWW.ONTHECOMMONS.ORG

The Farm Bill
and the Rise and Fall of Industrial Agriculture
Part 1: The Rise

Before the industrial revolution, farms were...

biodiverse

small-scale

largely self-sufficient

closed nutrient cycles

- solar-powered/ carbon-neutral (because people + animals, who do the work, get their energy from plants (o, animals who do) which get their energy from the sun

CO_2

Then came...

TRACTORS

now one person can farm more land!

soil erosion + depletion

cash cropping monoculture

The farm bill originated as a piece of FDR's 'New Deal' to help fix the Great Depression + Dust Bowl. It did 3 things:

the Great Depression +

DUST BOWL

So then came...

The Farm Bill

2. Soil conservation
New Deal farm bill programs paid farmers to set land aside — based on the cost of production

market flooding

lower prices

need to grow more to earn $

Then came...

1. 'Ever-Normal Granary'
the original farm bills managed **supply** by setting up grain reserves - buying extra in bountiful years + selling it back in lean ones. these reserves also provided early food aid

3. Price floors
established minimum prices based on the cost of production

with new international markets absorbing surplus + raising prices, price floors were eliminated. then prices dropped, and **subsidies** were implemented to keep farming low, benefiting processors

SYNTHETIC agricultural CHEMICALS
(pesticides, herbicides, fertilizers)
(which further boost production)

$956.4 total billion

Farm Bill spending today

Nutrition - Food Stamps/EBT... 11.1%

commodity subsidies
Crop insurance
conservation energy + resource + polluting
everything else

9.4% 4.6% 6% .1%

Then came...

INTERNATIONAL TRADE
fuels climate change
pesticides poison people + wildlife

synthesized fertilizers run off into waterways, into waterways dead creating dead zones

which also keep prices low. then food is

transported and processed

JUNK FOOD

end result: INDUSTRIAL FARMING as the dominant paradigm

The Farm Bill
and the Rise and Fall of Industrial Agriculture
Part 2: The Fall

Every 5-7 years, the farm bill expires & congress must pass a new one.

Here's what they do:

Here's what you can do:

Ideas for a better farm bill

The process begins with field & DC hearings. Stakeholders & state legislators share their priorities with their for the bill.

Then, House & Senate Agriculture Committees each draft, debate, and vote to pass a farm bill—resulting in **two bills**—each of which then goes to the 'floor' (the full legislative body) for more debate, amendments, and a **vote**

Then the two farm bills face off in 'conference committee,' where they are bundled into a single bill, which then goes back to House & Senate floors for one last vote and, once passed goes to the White House for the President to sign (or veto)

SUSTAINABLE FARMING

Hold your own hearing! Invite your legislator to tour your farm!

FAIR FOOD

Get media attention for your priorities with a cool event!

Write letters to the editor of your local newspaper!

call your legislators!

1 (202) 224-3121 capitol switchboard

GOT IT

Make use of the good farm bill programs to farm well!

grow solutions!

Replace wasteful subsidies with price floors

Address market concentration with fair competition laws

Rebuild local food infrastructure and rural communities

Support sustainable farming

Ensure access to healthy food

wild areas

diversified crops

animal & human powered

provides healthy food to the local community

animals grazing

hoop house

season extending

Ariana Taylor-Stanley

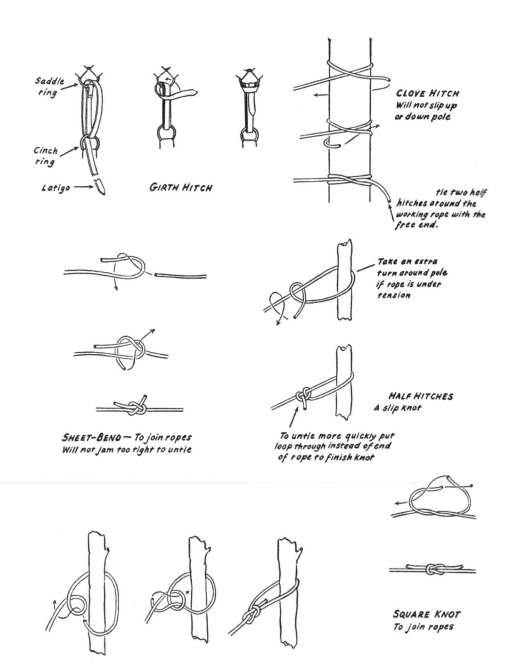

Saddle ring

Cinch ring

Latigo

GIRTH HITCH

CLOVE HITCH
Will not slip up
or down pole

tie two half
hitches around the
working rope with the
free end.

Take an extra
turn around pole
if rope is under
tension

SHEET-BEND ~ To join ropes
Will not jam too tight to untie

HALF HITCHES
A slip knot

To untie more quickly put
loop through instead of end
of rope to finish knot

BOWLINE ~ For a loop that will not slip

SQUARE KNOT
To join ropes

From Uncivilization:
The Dark Mountain Manifesto

Dougald Hine & Paul Kingsnorth

…The last taboo is the myth of civilisation. It is built upon the stories we have constructed about our genius, our indestructibility, our manifest destiny as a chosen species. It is where our vision and our self-belief intertwine with our reckless refusal to face the reality of our position on this Earth. The two are intimately linked. We believe they must be decoupled if anything is to remain. We believe that artists—which is to us the most welcoming of words, taking under its wing writers of all kinds, painters, musicians, sculptors, poets, designers, creators, makers of things, dreamers of dreams—have a responsibility to begin the process of decoupling. We believe that, in the age of ecocide, the last taboo must be broken—and that only artists can do it.

This response we call Uncivilised art, and we are interested in one branch of it in particular: Uncivilised writing. Uncivilised writing is writing which attempts to stand outside the human bubble and see us as we are: highly evolved apes with an array of talents and abilities which we are unleashing without sufficient thought, control, compassion or intelligence. Against the civilising project, which has become the progenitor of ecocide, Uncivilised writing offers not a non-human perspective—we remain human and, even now, are not quite ashamed—but a perspective which sees us as one strand of a web rather than as the first palanquin in a glorious procession.

TO THE FOOTHILLS!

A movement needs a beginning. An expedition needs a base camp. A project needs a headquarters. Uncivilisation is our project, and the promotion of Uncivilised writing—and art—needs a base. We present this manifesto not simply because we have something to say—who doesn't?—but because we have something to do. We hope to create a spark. If we are successful, we have a responsibility to fan the flames. This is what we intend to do. But we can't do it alone.

This is a moment to ask deep questions and to ask them urgently. All around us, shifts are under way which suggest that our whole way of living is already passing into history. It is time to look for new paths and new stories, ones that can lead us through the end of the world as we know it and out the other side. We suspect that by questioning the foundations of civilisation, the myth of human centrality, our imagined isolation, we may find the beginning of such paths.

If we are right, it will be necessary to go literally beyond the Pale, outside the stockades we have built—the city walls, the original marker in stone or wood that first separated "man" from "nature." Beyond the gates, out into the wilderness, is where we are headed. And there we shall make for the higher ground for, as Robinson Jeffers wrote, "when the cities lie at the monster's feet / There are left the mountains." We shall make the pilgrimage to the poet's Dark Mountain, to the great, immovable, inhuman heights which were here before us and will be here after, and from their slopes we shall look back upon the pinprick lights of the distant cities and gain perspective on who we are and what we have become.

This is the Dark Mountain project. It starts here.

Where will it end? Nobody knows. Where will it lead? We are not sure. Its first incarnation, launched alongside this manifesto, is a website, which points the way to the ranges. It will contain thoughts, scribblings, jottings, ideas; it will work up the project of Uncivilisation, and invite all comers to join the discussion.

Then it will become a physical object, because virtual reality is, ultimately, no reality at all. It will become a journal, of paper, card, paint and print; of ideas, thoughts, observations, mumblings; new stories which will help to define the project, school, movement of Uncivilised writing. It will collect the words and the

images of those who consider themselves Uncivilised and have something to say about it; who want to help us attack the citadels. It will be a thing of beauty for the eye and for the heart and for the mind, for we are unfashionable enough to believe that beauty—like truth—not only exists, but still matters.

Beyond that… all is currently hidden from view. It is a long way across the plains, and things become obscured by distance. There are great white spaces on this map still. The civilised would fill them in; we are not so sure we want to. But we cannot resist exploring them, navigating by rumours and by the stars. We don't know quite what we will find. We are slightly nervous. But we will not turn back, for we believe that something enormous may be out there, waiting to meet us.

Uncivilisation, like civilisation, is not something that can be created alone. Climbing the Dark Mountain cannot be a solitary exercise. We need bearers, sherpas, guides, fellow adventurers. We need to rope ourselves together for safety. At present, our form is loose and nebulous. It will firm itself up as we climb. Like the best writing, we need to be shaped by the ground beneath our feet, and what we become will be shaped, at least in part, by what we find on our journey.

If you would like to climb at least some of the way with us, we would like to hear from you. We feel sure there are others out there who would relish joining us on this expedition.

Come. Join us. We leave at dawn.

THE EIGHT PRINCIPLES OF UNCIVILISATION

"We must unhumanise our views a little, and become confident as the rock and ocean that we were made from." —Robinson Jeffers

Josef Beery

1

We live in a time of social, economic and ecological unravelling. All around us are signs that our whole way of living is already passing into history. We will face this reality honestly and learn how to live with it.

2

We reject the faith which holds that the converging crises of our times can be reduced to a set of "problems" in need of technological or political "solutions."

3

We believe that the roots of these crises lie in the stories we have been telling ourselves. We intend to challenge the stories which underpin our civilisation: the myth of progress, the myth of human centrality, and the myth of our separation from "nature." These myths are more dangerous for the fact that we have forgotten they are myths.

4

We will reassert the role of storytelling as more than mere entertainment. It is through stories that we weave reality.

5

Humans are not the point and purpose of the planet. Our art will begin with the attempt to step outside the human bubble. By careful attention, we will reengage with the non-human world.

6

We will celebrate writing and art which is grounded in a sense of place and of time. Our literature has been dominated for too long by those who inhabit the cosmopolitan citadels.

7

We will not lose ourselves in the elaboration of theories or ideologies. Our words will be elemental. We write with dirt under our fingernails.

8

The end of the world as we know it is not the end of the world full stop. Together, we will find the hope beyond hope, the paths which lead to the unknown world ahead of us.

Uncivilisation: The Dark Mountain Manifesto was published in 2009. Since then, the Dark Mountain Project has grown into an international network of writers and thinkers, artists and makers, farmers and foragers. So far, the fruits of this network have included festivals and gatherings, books, albums and conversations. At its heart is the Dark Mountain journal, a series of hardback books full of new writing and images from people who have taken up the challenge of the manifesto.

dark-mountain.net/about/manifesto

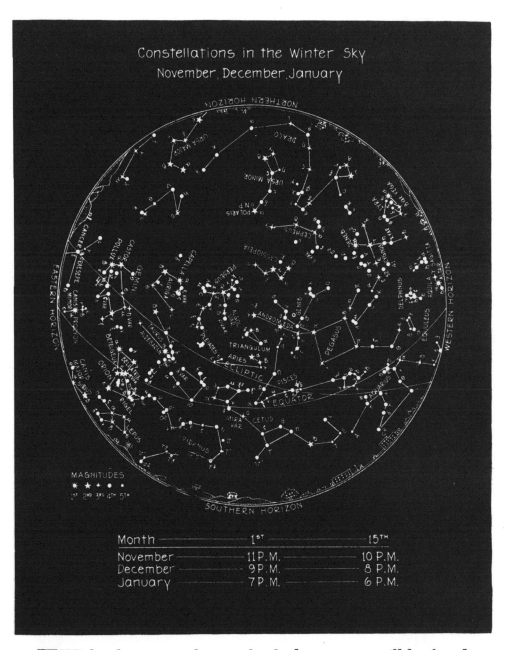

Constellations in the Winter Sky
November, December, January

Month	1ST	15TH
November	11 P.M.	10 P.M.
December	9 P.M.	8 P.M.
January	7 P.M.	6 P.M.

THE brightest stars have individual names, as will be found from the charts. Since early times the stars in each constellation have been known by Greek letters, in order of magnitude, the brightest star of the constellation being a (Alpha), the next brightest b (Beta).

303

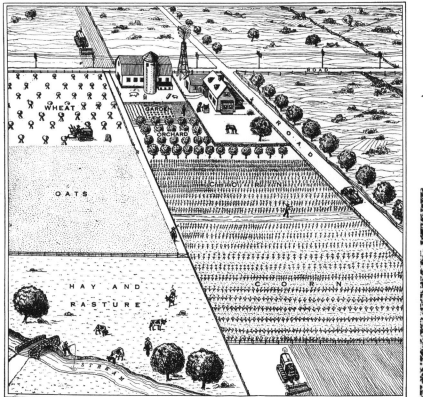

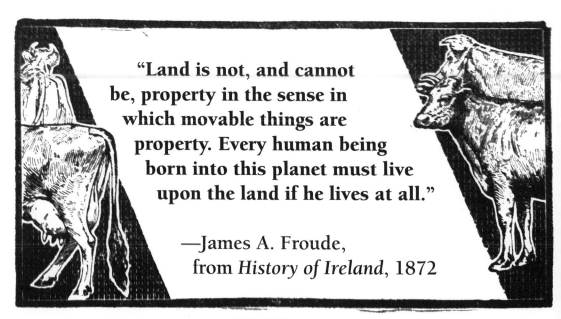

"Land is not, and cannot be, property in the sense in which movable things are property. Every human being born into this planet must live upon the land if he lives at all."

—James A. Froude, from *History of Ireland*, 1872

A Draft Agenda For Fixing The Food System

Liz Brownlee

Grab your red pen, Greenhorn, I need you to mark up this draft. It's an agenda for fixing our food system, and it's imperfect. I'm asking you to critique it so that we can clarify our ideas and build a better agenda together.

Some items are sweeping and expansive. They all will take time and sweat. Cross out every idea does not seem worth the costs. Then let your pen sail across the margins. Fill in missing agenda items and examples.

Flesh out ideas I neglected. There's a drawing exercise, too. Rearrange my food system map or draw your own to show the pieces, connections, and leverage points we can use moving forward.

Tonight I'll check my faded black mailbox (and my inbox) for your feedback. I'll incorporate your ideas and publish the improved, cross-pollinated draft. We'll be another step closer to changing our reality.

A PRACTICAL DESIGN to Bridge the gap!

Ginny Maki

AGENDA ITEM	NATIONAL OR POLICY ACTION	LOCAL OR COLLECTIVE ACTION	INDIVIDUAL OR COLLECTIVE ACTION
Make good food more affordable and accessible.	Double food stamps' value at farmer's markets.	Modernize city and town policy so that neighbors can easily create community gardens.	Challenge your regular customers to grow their own greens and zucchini (prolific and hard to kill crops). Sell starts to jump-start their efforts and make the challenge profitable.
	Challenge Americans to plant Victory gardens and community gardens.		Grow nutrient-dense foods like redleaf, open headed lettuce. It offers more anti-cancer agents than green or head varieties. Jo Robinson's *Eating on the Wild Side* has ideas.
Embark on an education campaign that gets people growing and eating good food.	Incentivize schools, nursing homes, and other institutions to create gardens, kitchens, and cooking classes.	Support local educators' efforts to connect kids to their food. Example: reimagine physical education (break a sweat while raising chickens on pasture), home economics (learn the thrift of parting a pasture-raised chicken), and health class (understand healthy portions).	
	Require school cafeterias to cook whole foods. The revolution in San Francisco cafeterias offers inspiration.	Figure out how to supply local institutions with reasonably priced food from local farms. Ask successful Farm-to School programs for advice.	
	Reward health insurance policy holders for eating healthy diets.	Challenge local journalists and artists to depict the changing face of food.	
	Fund programs that help beginning farmers build economically viable small businesses.		
	Support programs that retrain the workforce to grow, raise, breed, slaughter, butcher, distribute, sell, cook, and serve good food.	Define local food as regionally unique, sustainable, inclusive, and important. Kentucky and Maine provide good examples.	Host events on your farm or with your food. Think barn dances, pie-eating contests, watermelon festivals, and bicycle tours of area farms.

AGENDA ITEM	NATIONAL OR POLICY ACTION	LOCAL OR COLLECTIVE ACTION	INDIVIDUAL OR COLLECTIVE ACTION
Improve local and national food policies	Pass a 50-year Farm Bill. Incentivize sustainable agriculture, on-farm conservation practices, and good nutrition. End subsidies to conventional agriculture while we're at it. Wendell Berry's essay on the Farm Bill has more ideas. Label those gosh-darn GMOs.	Pass "Right to Farm" or other zoning regulations in your town or state to protect your right to farm in the face of urban sprawl. Michigan legislation should provide guidance.	
Scale up sustainable food production to meet our country's food demands.	Encourage the development of small farms, co-ops, food hubs, and local food restaurants via technical assistance, tax breaks, or other financial incentives.	Share your tool, system, business, or other innovations.	
	Expand training opportunities for conventional farmers transitioning to sustainable agriculture.	Collaborate with area growers to reach regional markets via collaborative CSA's, food hubs, and regional distributers that reach urban areas.	
Conserve our natural resources and adapt to a changing climate.	Pass climate change legislation. Perhaps:	Create local funding sources for conservation projects and climate adaptation. The ECOpass program Oklahoma is a good model.	Implement any best management practices you're missing (plant riparian buffers, use cover crops, and graze rotationally).
	Fully fund on-farm conservation programming through the Farm Bill.		Plant 100 fruit or nut trees or trial an agroforestry practice on your farm (like windbreaks, alley cropping, or forest farming). Either way, you'll capture carbon, diversify your crops, and build your flood and drought resiliency.
Conserve our natural resources and adapt to a changing climate.	Create a robust carbon market serving large and small forest owners.		
	Pay or incentivize farmers to provide critical ecosystem services like flood absorption, pollination, soil-carbon sequestration, and water filtration.		Trial crops from a warmer hardiness zone to take advantage of climate change.

AGENDA ITEM	NATIONAL OR POLICY ACTION	LOCAL OR COLLECTIVE ACTION	INDIVIDUAL OR COLLECTIVE ACTION
Understand the food system and spread the word.	Fund university research that asks questions like:	Collaborate with nonprofits to ask local food system questions.	Tell your food system story—it just may shift someone's thinking. Consider writing about how farming impacts:
	Can the US really feed itself? How many more jobs can the US support if we switched to mostly small, sustainable farms?		Your (or your community's) food sovereignty, the local economy, health care, or the natural world. Use outlets like your local public radio station, newspaper, or extension newsletter.
Write in your own ideas here:			

Nature As Measure:
Wes Jackson And The Land Institute

Megan Connolly

"Ecology is our primary field of interest because nature is our standard, the model we use as we design our experiments."
—Wes Jackson, Becoming Native to This Place

The Land Institute, located in Salina, Kansas, is dedicated to the idea that "a solution for the 10,000 year-old problem of agriculture"—soil degradation, ecosystem destruction, and high energy use—is necessary to ensure food security and biological integrity. Wes Jackson, founder and president of the Institute and a Kansas native, states that "the conflict between humans and nature is apparent, for agriculture is primarily production oriented, while nature's emphasis is upon preserving potential.

For nature, production need only be sufficient in order to ensure that potential is preserved. Humans reward enterprise, while nature rewards patience." It is the goal of the Institute to change agriculture from extractive and damaging to restorative and nurturing. To achieve the resilience of natural ecosystems, perennial grain polycultures must be brought to farms.

The majority of the food consumed by humans comes from annual grains, legumes, and oil seed crops—about 70 percent of global agricultural land is dedicated to producing them. Annual monocultures depend heavily on pesticides, commercial fertilizer and fossil fuel. The monocultured landscape creates the grounds for a growing number of pathogens and insects, controlled by a growing number of toxic pesticides.

The prairie, on the other hand, counts on diversity of species and genetics to avoid epidemics of insects and pathogens. The prairie maintains its own fertility, runs on sunlight, and actually accumulates ecological capital as soil. This accumulated soil stays in perennial pastures, native prairie grasslands, and forests without human action (Jackson, 2006).

The Land Institute is investigating ways to grow perennial grains, oilseeds and legumes together. These perennial crops anchor soil better and are less dependent on nitrogen-based fertilizers. They also manage pathogens and pests naturally, provide food for years without replanting, and sequester carbon at a smaller energy cost. Additionally the deep roots of perennial grains better handle the droughts or heavy rains that are likely to escort climate change.

The Institute primarily works from native ecosystems to create functional agroecosystems, but they also breed perennial grains that don't currently exist. These crops are bred in one of two ways: either with wild perennials selected for the best crop potential (domestication), or crosses between an annual grain and a related perennial species (hybridization). The Institute's scientists look at yield, seed size, seed retention, plant height and other qualities that contribute to a steady, substantial harvest the most promising plants are crossed so that each generation has a better chance at the desired traits. In Kansas, Jackson and his researchers are working on four main crops: intermediate wheatgrass (Kernza), wheat, sunflowers, and sorghum (The Land Institute, 2013).

For Jackson and the Institute the goals are simple: no soil erosion, no chemical contamination of the countryside, no fossil fuel dependency, a return to the land and to small towns and rural communities.

Sources

Jackson, Wes. (1994). *Becoming Native to this Place.* Lexington, Kentucky: University Press of Kentucky.

The Land Institute. (2013). What We Do and Why. Retrieved August 5, 2013, from http://www.landinstitute.org/vnews/display.v/ART/2013/03/06/513a22329f471.

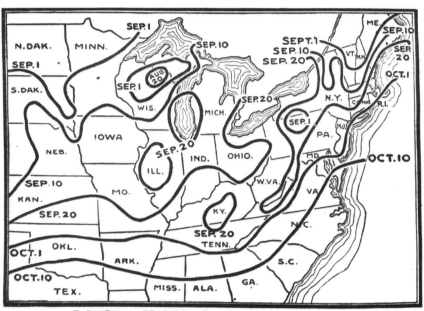

Earliest Dates on Which Killing Frosts Have Occurred

We came poor people to a seemingly empty land that was rich in resources. And based on that perception of reality—"poor people," "seemingly empty land," "rich"—we built our political, educational, economic, and religious institutions. Now we've become rich people in an increasingly poor land that's filling up, and the old institutions don't hold. So here we are. We patch things up, give them a lick and a promise, and things don't quite work. –Dan Luten, as quoted by Wes Jackson

MARCH—A SONG.

THE WORDS AND MUSIC COMPOSED FOR MERRY'S MUSEUM.

March is like a rill,
Now roaring, and now still;
To-day the blast is stinging,
To-morrow birds are singing.

March is like a cloud,
Now bright, and now a shroud;

To-day the warm rain falls,
To-morrow we have squalls.

March is like a bear,
With sharp claws and soft hair;
To-day 'tis rough and wild,
To-morrow, all is mild.

Songbook

1. Where There's a Will
2. When a Hundred Years Have Rolled
3. Tis Better to Stay on the Farm
4. The Plow, Spade & Hoe
5. The Hand That Holds The Bread
6. Keep Politics off Your Farm
7. Uncle Sam's Farm
8. Sink 'Em Low*
9. Diamond Joe*
10. Do Not Mortgage the Farm
11. Work (Our Little Grangers)

All songs collected by Brian Dewan,
except *songs collected by Max Godfrey.

Where There's A Will

Charles Edward Pollock

Tho' troubles perplex you,
Dishearten and vex you,
Retarding your progress in somber array;
To shrink from with terror
Is surely and error,
For where there's a will there is always a way.

CHORUS:
There's a way, there's a way,
Wherever there's a will there's a way.
There's a way, there's a way,
Wherever there's a will there's a way.

The task may be teasing,
The duty unpleasing,
But he who confronts it will soon win the day;
The fight is half over
When once we discover
That where there's a will there is always a way.

CHORUS

Misfortunes uncounted
Are often surmounted,
If only we quit not the field in dismay;
Then one more endeavor,
Remembering ever,
That where there's a will there is always a way.

CHORUS

Orr, James L., compiler; Pollock, Charles Edward. "Where There's a Will There's a Way" in *Grange Melodies*, published by the National Grange of Patrons of Husbandry, for Use in the Granges of the United States," Philadelphia, PA: Geo. S. Ferguson Co, 1911. Permission: Northern Illinois University

When A Hundred Years Have Rolled

Charles H. Gabriel

Oh, where will be the things I love,
When a hundred years have rolled?
Oh, who will roam this garden then,
And claim this shining gold?
Where will these little flowers be,
That bloom so bright so fragrantly?
These little birds, so blithe and free,
When a hundred years have rolled?

CHORUS:
When a hundred years have rolled,
When a hundred years have rolled,
Oh, where will be the things I love
When a hundred years have rolled?
Oh, where will be my new-found friends,
When a hundred years have rolled?

These hearts that hope to me extend,
Though feeble now and old;
Their hoary heads are bending low,
They long have stood earth's pain and woe.
They'll not be grov'ling here below.
When a hundred years have rolled.

CHORUS

Where will these little children be,
When a hundred years have rolled?
That now are happy, glad, and free,
That fondly now we fold;
Oh, where, my soul, where wilt thou be?
Does hope of heav'n extend to thee?
Prepare, prepare with God to be
When a hundred years have rolled.

CHORUS

Orr, James L., compiler; Gabriel, Charles H. "When a Hundred Years Have Rolled" in *Grange Melodies*, published by the National Grange of Patrons of Husbandry, for Use in the Granges of the United States, Philadelphia, PA: Geo. S. Ferguson Co, 1911. Permission: Northern Illinois University

'Tis Better To Stay On The Farm

E.R. Latta & J.H. Tenne

As guiding his plow 'mid the corn rows,
The grass and the weeds to destroy,
What wonderful day dreams of pleasure
Take form in the mind of the boy.
He knows not the wiles of the city,
So frequently leading to harm;
My boy, from your reverie waken,
'Tis better to stay on the farm.

CHORUS:
'Tis better to stay on the farm, my boy,
'Tis better to stay on the farm;
Your dreaming will lead you to harm, my boy,
Will certainly lead you to harm;
'Tis better to stay on the farm.

The gas-lighted hall, with its pleasures,
He dreams of, and longs to be there;
And heedless of trouble and labor,
He thither-ward seems to repair.
"How stupid a life in the country,
The city has many a charm!"
My boy, from your reverie waken,
'Tis better to stay on the farm.

CHORUS

He dreams he's a clerk at the counter,
And thinks if it almost divine;
He toils in the cornfield no longer,
No more shall his spirit repine.
But hearken! the noon bell is calling,
The dreamer starts up in alarm!
My boy, if you'll only believe me,
'Tis better to stay on the farm.

CHORUS

The Plow, Spade & Hoe

The farmer is chief of the nation,
The oldest of nobles is he;
How blest beyond others his station,
From want and from envy how free!
His patent was granted in Eden,
Long ages and ages ago;

CHORUS:
Oh, the farmer, the farmer forever,
Three cheers for the plow, spade, and hoe.

In April, when nature is waking,
And bluebirds are first on the wing,
His plow now the fallows are breaking,
Whence beautiful harvest shall spring;
Then broadcast along the brown furrow,
We hasten the good seed to sow;

CHORUS

But when, in the clear Autumn weather,
He reaps the reward of his care;
So busy and joyful together,
What monarch with him can compare?
His barns running over with plenty,
His trees with their fruit bending low;

CHORUS

Then sing me the life of a farmer,
With comfort and health in his train,
And heed not the voice of the charmer,
That whispers of speedier gain;
With all the rich treasures 'tis teeming,
That heaven on man can bestow;

CHORUS

The Hand That Holds The Bread

George Frederick Root

Brothers of the plow! The power is with you;
The world in expectation waits for action prompt and true,
Oppression stalks abroad, monopolies abound;
Their giant hands already clutch the tillers of the ground.

CHORUS:
Awake, then, awake! The great world must be fed,
And Heaven gives the power to the hand that holds the bread,
Yes! Brothers of the plow! The people must be fed,
And Heaven gives the power to the hand that holds the bread.

Brothers of the plow! In calm and quiet might,
You've waited long and patiently for what was yours by right;
A fair reward for toil, a free and open field;
An honest share for wife and home of what your harvests yield.

CHORUS

Brothers of the plow! Come rally once again,
Come, gather from the prairie wide, the hillside and the plain;
Not as in days of yore, with trump of battle's sound,
But come and make the world respect the tillers of the ground.

CHORUS

Keep Politics Off Your Farm

Laura E. Newell & Charles Edward Pollock

Some curious weeds I might mention
That lend to the landscape no charm;
To one let me call your attention,
Keep politics off of your farm.

Tho' weeds will with politics mingle,
Potatoes with politics fail;
Devote your whole mind to your business,
And make ev'ry effort avail.

CHORUS:
Keep politics off of your farm
Your crops they well certainly harm;
If you would successfully labor,
Keep politics off of your farm.

Just keep an eye open to business,
Keep posted, but stick to your text;
Don't be disconcerted by trifles,
And don't be too easily vexed.

Don't spend all your time riding hobbies,
Predicting distress and alarm;
You'll find it a great disadvantage
To grow politics on your farm.

CHORUS

Oh, this is an age of advancement,
And kings are the "sons of the soil;"
But political schemers and "bosses"
Full many an effort will foil.

Pray, take this advice as a warning,
You'll find it will work like a charm:
Apply yourself strictly to business,
Keep politics off of your farm.

Uncle Sam's Farm

Jesse Hutchinson Jr. & N. Barker

Of all the mighty nations in the East or in the West,
O this glorious Yankee nation is the greatest and the best.
We have room for all creation and our banner is unfurled,
Here's a general invitation to the people of the world.

CHORUS:
Then come along,
come along, make no delay;
Come from every nation,
come from every way.
Our lands, they are broad
enough—don't be alarmed,
For Uncle Sam is rich enough
to give us all a farm.

St. Lawrence marks our
Northern line
as fast her waters flow;
And the Rio Grande
our Southern bound,
way down to Mexico.
From the great
Atlantic Ocean
where the sun
begins to dawn,
Leap across the Rocky Mountains
far away to Oregon.

CHORUS

While the South
shall raise the cotton,
and the West, the corn and pork,
New England manufactories
shall do up the finer work;
For the deep and
flowing waterfalls
that course along our hills
Are just the thing for
washing sheep
and driving cotton mills.

CHORUS

Our fathers gave us liberty,
but little did they dream
The grand results that pour
along this mighty age of steam;
For our mountains,
lakes and rivers are all
a blaze of fire,
And we send our news
by lightning
on the telegraphic wires.

CHORUS

The brave in every nation
are joining heart and hand
And flocking to America,
the real promised land;
And Uncle Sam stands ready
with a child upon each arm
To give them all a welcome
to a lot upon his farm.

CHORUS

A welcome, warm and hearty,
do we give the sons of toil
To come to the West
and settle and labor on free soil;
We've room enough
and land enough,
they needn't feel alarm—
O! come to the land of freedom
and vote yourself a farm.

CHORUS

Yes! we're bound to lead the
nations for our motto's "Go
ahead,"
And we'll tell the foreign paupers
that our people are well fed;
For the nations must remember
that Uncle Sam is not a fool,
For the people do the voting
and the children go to school.

CHORUS

Sink 'Em Low

Submitted By Max Godfrey

Well if you wanna,
Please your captain,
Sink 'em low boys,
Raise 'em high.

CHORUS:
Sink 'em low boys,
Sink 'em low,
Sink 'em low, boys,
Raise 'em high

I asked the judge,
What may be my fine, boys,
Said if I don't hang you,
Give you ninety-nine.
Give you ninety-nine, boys,
Give you ninety-nine,
Said if I don't hang you,
Give you ninety-nine.

CHORUS

I asked y captain,
Has Saturday morning come,
boys,

Said it makes no difference,
I don't owe you none.
I don't owe you none, boys,
I don't owe you none,
Said it makes no difference,
I don't owe you none.

CHORUS

If I could just make it,
Through January
and February,
I'd march on through, boys,
And I'd go back home.
And I'd go back home, boys,
I'd go back home,
I'd march on through, boys,
I'd go back home.

CHORUS

I asked that captain,
What's the time of day, boys,
He was so hard-hearted,
He just walked away.

He just walked away, boys,
Just walked away,
He was so hard-hearted,
He just walked away.

CHORUS

I thought I heard my,
Big-leg Lula,
She callin' me, boys,
Sayin' come on home.
Sayin' come on home, boys,
Come on home,
I heard big-leg Lula,
Sayin' come on home.

CHORUS

So if you wanna,
Please your captain,
Sink 'em low, boys,
Raise 'em high.

SOURCE Recorded in 1960 on St. Simons Island, sung by the Georgia Sea Island Singers, consisting of Bessie Jones, John Davis, Peter Davis, Willis Proctor, and Henry Morrison. According to the Association for Cultural Equity, "The unidentified handwriting on tape box identifies this as a 'shoveling song' (sung by the men who dug highway trenches). The men in the chorus provide grunts and aspirations at the middle and end of each line. John Davis exclaims 'I don't want no captain like that!' at end."

Diamond Joe

Submitted By Max Godfrey

Diamond Joe, come-a git me,
Diamond Joe, come-a git me,
Diamond Joe, come-a git me,
Diamond Joe.

Diamond Joe, come-a git me,
Diamond Joe, come-a git me,
Diamond Joe, come-a git me,
Diamond Joe.

REFRAIN:
Well I went up on the mountain,
And I give my horn a blow,
Thought I heard my Maybelle say,
Yonder come my beau,

Well I ain't goin' work in the factory,
And I can't afford a farm,
I'm goin' wait til my Maybelle come,
And she goin' call me home.

REFRAIN

Well I ain't goin' tell you no story,
And neither would I lie,
I'm goin' wait til my Maybelle come,
And far away we"ll fly.

REFRAIN

I went up Kennesaw mountain,
And I gave my horn a blow,
Prettiest girl in Atlanta come,
Walkin' up to my door.

Diamond Joe, where'd you find him?
Diamond Joe, where'd you find him?
Diamond Joe, where'd you find him?
Diamond Joe.

Diamond Joe, where'd you find him?
Diamond Joe, where'd you find him?
Diamond Joe, where'd you find him?
Diamond Joe.

Most of the verses are from Charlie Butler's version, which can be found on the Alan Lomax collection *Deep River of Song–Mississippi: Saints and Sinners*. The verse about Kennesaw mountain is from "Atlanta Strut" by Blind Willie McTell. I've slightly changed the verses from the recordings I originally learned them from. A transcription of Charlie Butler's original verses can be found in Stephen Wade's great new book, *The Beautiful Music All Around Us*. Recordings of me singing "Diamond Joe" can be found on the Intergalactic Agrarian Mixtape as well as on my album *Down Don't Worry Me*.

Do Not Mortgage The Farm

E.R. Latta & James L. Orr

Fortune may sometimes forsake you,
Useless the struggle may seem;
But be not tempted to hazard
That which you may not redeem;
Do not imperil the homestead,
Banish the thought in alarm,
Make it your strong resolution,
Never to mortgage the farm.

CHORUS:
Do not mortgage, not mortgage the farm,
Do not mortgage, not mortgage the farm;
For sorrow will soon over take you
If ever you mortgage the farm.

Think of the time it has taken,
Think of the toil it has cost,
That you and your children might own it,
Now do not let it be lost;
Think of the hearts that enshrine you,
And trust you to shield them from harm,
Make it your strong resolution,
Never to mortgage the farm.

CHORUS

If you would peacefully slumber,
Knowing no waking regret,
See that your right to the homestead
Is not encumbered by debt;
Strictest economy practice,
Toil with a vigorous arm,
Make it your strong resolution,
Never to mortgage the farm.

CHORUS

"Work" (Our Little Grangers)

James L. Orr

Don't think that there's nothing for children to do
Because they can't work like a man;
The harves is great and the lab'rers are few,
Then, children, do all that you can.

CHORUS:
Then work, work, work, children, work,
There's work for the children to do,
Then work, work, work, children, work,
There's work for the children to do.

You think if great riches you had at command,
Your zeal should no weariness know;
You'd scatter your wealth with a liberal hand,
And succor the children of woe.

CHORUS

But what if you've nough but a penny to give,
Then give it, tho' scanty your store;
For those who give nothing when little they have,
When wealthy will give little more.

CHORUS

**"Be not tedious in
discourse or in reading,
unless you find the
company pleased
therewith."**

—George Washington, *Rules of Civility*

326

327

Jardinage
Basse-Cour

Chasse et Pêche
Elevage

Sports
Bricolage

T.S.F.
Romans

RUSTICA

HEBDOMADAIRE ILLUSTRÉ

JOURNAL UNIVERSEL DE LA CAMPAGNE

N° 23
6 Juin
1937
—
Dixième année
32 PAGES

Lire dans ce numéro les articles :

**L'ENSACHAGE VOUS PERMETTRA D'OBTENIR
DE BEAUX ET BONS FRUITS**

LES CHENILLES TORDEUSES DES ARBRES FRUITIERS ET D'ORNEMENT
LES PLANTES GRIMPANTES A FLEURS OU A FEUILLAGE DÉCORATIF

60
CENTIMES

RÉDACTION ET ADMINISTRATION : 1, RUE GAZAN, PARIS (XIV°)

TELEPHONE
TREE
TEMPLATE

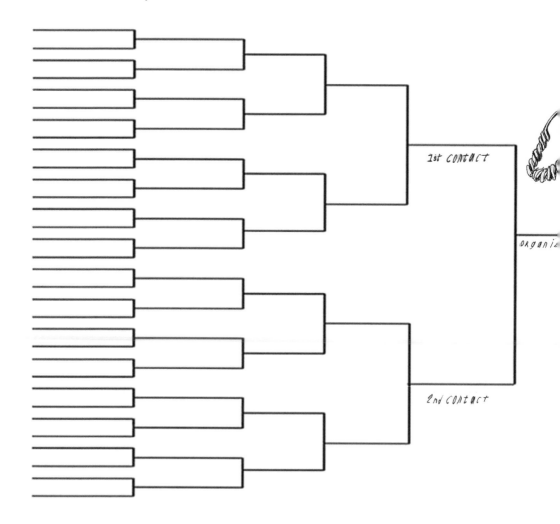

1st CONTACT

organiz

2nd CONTACT

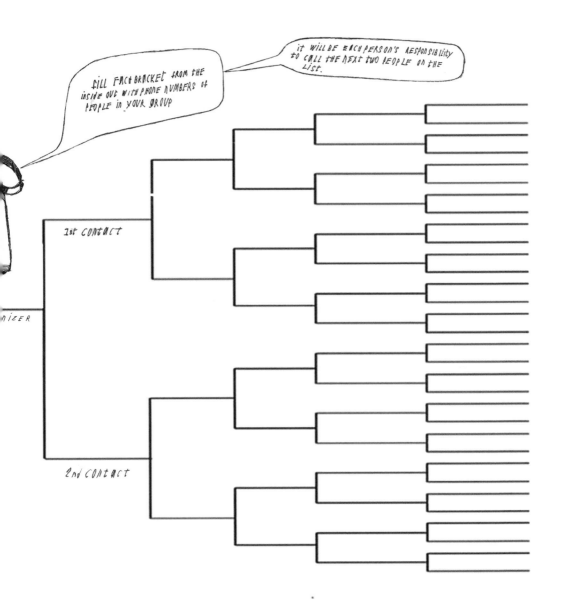

Brooke Budner

Intergalactic Agrarian Mixtape

We are pleased to present this open-source playlist for agrarians (also available on cassette tape and vinyl record).

The Intergalactic Agrarian Mixtape was compiled by Audrey Berman from many wonderful friends on both sides of the Atlantic, including, but not limited to: Brendan McMullen, Creek Iverson, Bennet Konesni, Edith Gawler, Max Godfrey, and Brian Dewan.

The songs have been recorded in granges, blizzards, festivals, and general agrarian splendor alike. It is also our first collaboration with the Landworkers Alliance.

Cassette tape and vinyl record versions are available from www.thegreenhorns.net, and the digital version can be downloaded here:

WWW.ARCHIVE.ORG/DETAILS/THENEWFARMERSALMANAC

Other Projects By The Greenhorns

Every year is a new year. And in 2015 much newness there is indeed. As you know, Greenhorns is a flexible network of edge-wild agrarians. We're geographically distributed, highly reactive and positively pulsating with new ideas that provoke change on the landscape, and economies it can feasibly sustain. Our mission hasn't changed, we are here to promote, support and recruit the entering generation of sustainable farmers. Our methods haven't changed: create innovative programing, publications, professional development materials, networking events, video, audio, art, literary and lively community happenings for those same farmers (and farmers-to-be). Yes, we are protean, ever improvising, and quite prolific.

We know that agriculture is culture too, that staring at the soil, the pasture, the piglet and the parsnip is a good time for deep thinking, and the right context to hatch a literary mid-life crisis-averting journal habit. This year, 2015 we're onto some great things, here's a little list to chew on:

Greenhorns Radio

Weekly podcast on Heritage Radio Network. Each week a new young or youngish farmer reflecting on the path, the dream, the logistics, the heartbreak. More than 200 episodes on archive so far:
http://www.heritageradionetwork.com/programs/7-Greenhorns-Radio

Intergalactic Agrarian Mix-Tape

A cassette tape (open source, creative commons) collaboration with the English Young Farmers movement, freely re-recordable, a set of songs about resistance. A project of Greenhorns & Farmworkers Alliance.

Maine Sail Freight Project

A follow up to 2013's maiden voyage of the Vermont Sail Freight (350 miles each way from Vermont to Manhattan with $60,000 of regional food) Our eventual goal with this one is a route from Lubec to Boston. This summer we'll pull off a highly choreographed sail-powered community potluck, cargo-logistics website and big-tent event to celebrate the various sail-freight projects in formation.

BIG Land (Benevolent Investors Guide)

Given the massive concentration of wealth in this country, and our funny new sub-club of billionaire land stewards, we decided to print a pro-active guidebook for wealthy people who want to invest in land and young farmers, and to help them do it right.

Posters

We're working on two posters, on Land Reform and Land Use Progression, one with Pat Perry about Ejido's, Sherman's Order, Terre de Liens, and dozens of other land reform movements around the world and through time. The other is a land-use history map of the Hudson-Champlain waterway, with Markley Boyer. Scroll forward to see the future.

The Land Magazine

a glorious occasional publication about land rights out of Devonshire, UK. Greenhorns is happy to publish and distribute it in the USA.

Our Land

web films about building a new farm economy, inside the old one. We tried hard to do them monthly, but really thats impossible. There are quite a lot of them still in the hopper—so stay tuned.

The New Farmer's Almanac

and its distribution along with our stickers, our posters, our small and large guidebooks, grange paraphernalia, t-shirts, tote bags etc.

Grange Future

is a multi-media project about the Grange movement, past and present, its core tenets, drama and revival, gathering oral histories,and testimony about community histories in 8 micro-regions around the country. We made a Grange Future travelling exhibit and have taken it around to Grange Halls. Have one in your town? Get in touch! Our first vinyl record "GRANGE FUTURE: Traditional Songs Sung by Brian Dewan," recorded in our active Adirondack grange halls.
WWW.GRANGEFUTURE.COM

NY State Lands Initiative

is a piece of legislation we need the Governor to sign. It puts NY State owned prison, municipal, transportation and mental health lands up for lease/ RFP for beginning farmers.

Serve Your Country Food Map

has been up for a while, but we've recently added a bunch of useful datasets, in particular agrarian lawyers and land trusts. www.serveyourcountryfood.net

Seed Circus 2016, 3-Day Bonanza

We're in pre planning mode for a serious festival celebrating the intersections and themes of the north country, and hyper rural young farmers movement. This includes draft power, seed saving, agro-forestry, farming with the wild, historic preservation, carbon farming, populism.

Save the Seed Campaign

Working to protect the human right to breed, save and share seed outside the commodity framework.

Chestnut Project

run by James Most who is building a west-coast nursery of grafted chestnut trees to distribute to farmers, ranchers, and landowners to promote crop diversification and land restoration with tree crops.

Soil Carbon

Along with Rebecca Burgess and Rio de La Vista, we're convening a stakeholder meeting on Soil Carbon at Paicines Ranch. There will be audio posted to Greenhorns.

Headquarters Mother Orchard

Eliza Greenman will lead the team installing a mother orchard for the north country at our brand-new headquarters in Westport NY. Join in the fun, come visit.

Seaweed Commons

A project inspired by the work of Nobel-prize winning economist Elinor Ostrum, to convene a council for commons-based natural resource management of edible seaweed along the coast of Maine. Public trainings and lectures to be held at College of the Atlantic, where they offer a degree in Human Ecology. This work will take us up to Alaska to study, speak, and learn with the Community Fisheries projects located there.

Land Reform Trainings

Greenhorns is partnering with Agrarian Trust to raise the profile of peaceful land reform, the history and promise of a powerful tactic for dis-aggregating.

Greenhorns Library and Archive

We've catalogued and open-sourced our library collection on LibraryThing. Soon we'll be shelving the books in our new headquarters in New York. Plenty of work for volunteers who love to read. Meanwhile, archivists at the Schumacher Center have begun the process of building an inter-library system of reading lists, excerpts and "self-training" modules on New Economics. We'll be pushing that along to you.

In these times: Rural Edition

As leadership begins to shift into an orchestral suite, Severine will be writing more. In These Times is a progressive magazine, Severine is pleased to join Winona Laduke as a founding editorial contributor.

Lost Landscapes: Rural Edition

Greenhorns is thrilled to be working in coalition with Rick Prelinger of the Prelinger Library on an "archival remix" following in his series of projects compiling home-video from San Francisco, New York, Detroit and other cities. These films are shown in a community context, they iterate, they learn, and it is the audience who provides the narration. We'll be taking this show on the road starting late summer 2015.

FFA Watchdog

Eliza Greenman led us off to a great start in 2014 when she brought a team down to the National Convention of Future Farmers of America in Kentucky. It's a group that gets federal as well as corporate funding to promote agricultural leadership. We, too, wish to promote agricultural leadership, but not as a "mouthpiece for ag-industry."

Land Justice

An ongoing investigation into durable land-holding configurations, many times in a spiritual / religious context. For now we're starting with research, on Catholic Worker farms, Navajo Reservations and Zen Monestaries.

Thrilling? Perplexing?

join the email listserv to get an essay every month, plus all the events, arcana, gossip and resource-dump we can muster.

www.thegreenhorns.net

Be part of the next Almanac

You may have noticed that we skipped a year
for this latest issue, and we thank you for your
patience. It's now our policy to solicit far in
advance.

All submissions for the 2017 New Farmer's
Alamanac are due February 1, 2016.

Send submissions and inquires to
almanac@thegreenhorns.net

www.thegreenhorns.net/category/media/almanac